When young Francis Egerton, the future 'Canal Duke' of Bridgewater, was making his Grand Tour of the Continent he made it his business to inspect the Canal du Midi. What he saw undoubtedly inspired his own pioneer Bridgewater Canal which heralded the great period of canal construction in England.

Francis was by no means the only eighteenth century nobleman to include the Canal du Midi in his itinerary, for it was then rightly regarded as one of the wonders of Europe. Built between 1666 and 1681, it was undoubtedly the greatest work of engineering in the world at that time. Nothing upon so ambitious a scale had been built by man since the fall of the Roman Empire.

Cutting across southern France from the westward-flowing Garonne at Toulouse to the port of Sete, its purpose was to provide an inland route for shipping between the Atlantic and the Mediterranean and thus avoid the long and hazardous sea passage via Gibraltar. One hundred and fifty miles long, and climbing by ladders of locks to a height of 600 feet, it was the first major summit level canal in the world. How to keep its summit adequately supplied with water in a southern climate where rainfall is low was the greatest problem its builders had to solve. Their successful solution consisted of forty miles of feeder channels and the construction of the mighty St Ferreol Dam.

Continued on back flap

BY THE SAME AUTHOR

The Aeronauts: A History of Ballooning

Waterways
Narrow Boat
Green and Silver
Inland Waterways of England
The Thames from Mouth to Source
Navigable Waterways

Railways
Lines of Character
Railway Adventure
Red for Danger

Motoring
Horseless Carriage
A Picture History of Motoring

Topography
Worcestershire

Philosophy
High Horse Riderless
The Clouded Mirror

Fiction
Sleep no More
Winterstoke

Biography
Isambard Kingdom Brunel
Thomas Telford
George and Robert Stephenson
The Cornish Giant (Richard Trevithick)
Great Engineers
James Watt
Thomas Newcomen

Engineering History
A Hunslet Hundred
Tools for the Job, a Short History of Machine Tools
Victorian Engineering

Autobiography
Landscape with Machines

L. T. C. Rolt

From Sea to Sea

The Canal du Midi

Allen Lane

Copyright © L. T. C. Rolt 1973

First published in 1973

Allen Lane
A Division of Penguin Books Ltd
21 John Street, London WC1

ISBN 0 7139 04712

Printed in Great Britain by
Western Printing Services Ltd, Bristol

Contents

List of Illustrations

ILLUSTRATION ACKNOWLEDGEMENTS

*Thanks are due to the following for supplying the photographs:
Sonia Rolt for Nos. 17, 18, 20, 21, 23, 24, 29, 30, 31, 32, 33,
34, 42. Richard Rolt for Nos. 4, 5, 6, 7, 8, 11. Timothy Rolt
for Nos. 3, 9. Michael Streat for Nos. 22, 43. Studio Yan,*

Toulouse for Nos. 1, 10, 13, 28, 37, 38, 39, 40, 41, 44. Nos. 2, 12, 14, 15, 16, 19, 25, 26 are taken from the book of plates accompanying Général Andreossy's Histoire du Canal du Midi *of 1804 and I acknowledge the courtesy of the Institution of Civil Engineers in allowing them to be photographed. The maps were drawn by Edward McAndrew Purcell.*

Introduction

In the early spring of 1752 a young English aristocrat left these shores for France to embark upon the Grand Tour. He was Francis Egerton, at this time a gauche, unprepossessing and sickly young man who drank too much. No one thought highly of his prospects, for his six elder brothers had all died of tuberculosis and he exhibited the same deadly symptoms. Yet he was destined to grow to vigorous manhood and to achieve permanent fame as the celebrated 'Canal Duke' of Bridgewater who, in association with the engineer James Brindley, and his Agent, John Gilbert, built the Bridgewater Canal. This canal, running from collieries on the Duke's estate at Worsley to Manchester and later extended to the Mersey at Runcorn, was entirely financed at considerable risk by the Duke. It was a great commercial success and a prototype which inaugurated the great age of canal construction in this country.

Acting as the young Duke's guide, philosopher and friend on his Grand Tour was Robert Wood, writer, traveller, archaeologist and polished man of the world who, on his return, became member of Parliament for the Duke's pocket borough of Brackley. While abroad, Wood wrote conscientiously to the Duke's uncle, the Duke of Bedford, reporting his protégé's health and progress. They were away for over three years. After a preliminary stay in Paris, they moved south to Lyons where the young Duke attended

the Academy. In the autumn of 1753, however, the Duke
expressed his desire to see the famous Languedoc Canal and,
having obtained permission from the Duke of Bedford, the
two set off by coach on this engineering pilgrimage. What
had quickened the Duke's interest in canals we do not know.
Some think that on an earlier excursion into Italy he may
have visited Milan and have seen those two pioneer fifteenth-
century Italian canals, the Martesana and the Bereguardo.
But his curiosity surely needs no explanation of this kind,
for the Languedoc Canal, Canal du Midi or Royal Canal, as it
was variously called, was undoubtedly the greatest work of
engineering in the world at that time. Nothing on so grand
a scale had been built by man since the fall of the Roman
Empire. No wonder that a young English nobleman
on the Grand Tour should wish to include it in his itinerary.

Built between 1666 and 1681, the purpose of the canal was
to provide an inland route for shipping from the Atlantic
to the Mediterranean, thus avoiding the long and hazardous
sea passage via Gibraltar. It runs for 149 miles in a south-
easterly direction across southern France from a junction
with the westward-flowing Garonne at Toulouse to the
Étang de Thau and the port of Sète on the coast of the
Mediterranean. From Lyons, the two travellers first visited
Aix-en-Provence and then travelled from there to Sète
where they explored the busy wharves and docks of a port
which the seventeenth-century canal engineers had created.
They then headed westwards, their heavy coach rumbling
slowly along the road that leads through Agde, Béziers,
Carcassonne and Castelnaudary to Toulouse. This road is
never very far away from the canal, for both follow the
natural divide that stretches across France between the

highlands of the Massif Central to the north and the snow-capped peaks of the Pyrenees.

Because the young Duke was an indifferent correspondent at this time, we have no means of knowing his reactions to what he saw. But we can guess that his imagination was fired by the great canal works he had seen because, on his return to the Lyons Academy, he insisted on attending a course in 'Experimental Philosophy', a term which then embraced both science and engineering. We know now what momentous consequences this journey had for England. For it was the impression made on him by his visit to the Canal du Midi (to use the name by which it is now best known) and the sense of wonder it had inspired in him that, soon after his return to Lancashire, encouraged him to embark on his own ambitious project that ushered in the canal age in England and so gave such tremendous impetus to Britain's Industrial Revolution.

This story explains how this book came to be written by a stay-at-home Anglophile with a minimal knowledge of France or the French language. My lifelong interest in canals inspired a strong desire to see for myself the Canal du Midi, that archetypal engineering work which, by proving that a summit level canal on the grand scale was practicable, set an example, not only to England, but to the world. Unlike the young Duke of Bridgewater, who only saw the Canal du Midi from the bank, I was able to travel through it by boat, which is the only right and proper way to get to know any canal. In the spring of 1971 I headed eastward from Castelnaudary as far as the little port of Marseillan on the Étang de Thau. At the same time I was able to explore the marvellous feeder system which supplies the summit level of the canal with water from the Montagne Noire. In the following September I travelled through the canal in the opposite direction, entering the canal from the Mediterranean through the port of Agde by the famous Round Lock, travelling to Toulouse and from thence down that later-built extension, the Canal Latéral à la Garonne, as far as Castets-en-Dorthe where it joins the tidal Garonne above Bordeaux.

These on-the-spot explorations and the historical research which accompanied them were a salutary experience. For approximately one hundred years, from 1760 to 1860, Britain led the world into the Industrial Revolution and this gave birth to the chauvinistic belief that British engineering skill was mysteriously and innately superior to that of any other nation. Despite all that has happened since, this view still lingers on. So, at a moment when Britain is about to join the European Community, my visit to the Canal du Midi was a timely reminder to me that, in the past no less than in the present, British engineers were not the only pebbles on the beach. For here I saw a work whose scale and grandeur is unmatched by anything that English canal engineers achieved a hundred years or more later. The moral is that engineering talent is no monopoly enjoyed by any particular nation, but that such talent cannot flower unless the particular social, political and economic climate is congenial. In England in the century after 1760 it was set fair for engineers of exceptional talent. Similarly, in seventeenth-century France, the Sun King smiled on Pierre Paul Riquet, the creator of the Canal du Midi.

Although the gauge of the Canal du Midi is small by modern continental standards it is considerably larger than most canals in England. Yet, at 620 ft above sea level, its summit level is only a mere 24 ft below that of the highest canal level in England, at Standedge on the Huddersfield Narrow Canal. This effectually disposes of another English myth which is that the commercial success of continental canals is entirely due to the fact that the continent is flat, allowing canals to be built with comparatively few locks, whereas England is a hilly country.

When I inquired about early historical works on the Canal du Midi I was told that two of them, published in France in 1778 and 1804 respectively, were now extremely scarce and that I might have to go to the Bibliothèque Nationale in Paris to consult them. To an impecunious author, this was a daunting prospect. Then I remembered that when Thomas Telford died, under the terms of his will his library passed into the possession of our Institution of Civil Engineers, of which he was the first President. I had a hunch that his collection would have included both these books and, sure enough, this proved to be correct. Not only did they bear Telford's signature, but also annotations in his handwriting in the margins to prove that here was one British canal engineer, at least, who took the trouble to study in a foreign tongue what his predecessors in France had accomplished. This led me to speculate why it was that our canals should seem so inadequate when compared to their famous French forerunner. It could be argued that the latter was intended to be a ship canal, whereas the much smaller dimensions of our canals were adequate to meet the needs of purely inland transport in eighteenth-century England. Yet the key to the success of the Canal du Midi, and its most outstanding feature, is the elaborate feeder system which ensures that its high summit level is adequately supplied with water. Having seen this system, it seems inexcusable that so many of our canals, though built in a country with a much higher and more regular rainfall, so frequently fall short in this essential respect. It can only be explained by the different circumstances in which they were built. Whereas the Canal du Midi enjoyed state patronage and no expense was spared to make it worthy of the reign of 'le Roi Soleil', with the sole exception of the Bridgewater, our canals were built by joint stock companies for purely commercial motives and their engineers had to work under managing committees who maintained a tight-fisted control over expenditure.

These committees, as our canal engineers were probably the first to realise, could often be penny wise and pound foolish. A good example of this was the rejection, purely on grounds of economy, of John Rennie's original plan for the summit level of the Kennet & Avon Canal. The cheaper plan finally adopted greatly reduced the length of the proposed summit tunnel at the price of a higher summit level which entailed constant pumping throughout the lifetime of the canal to keep it supplied with water. As Pierre Paul Riquet rightly pointed out as early as 1662, 'machines for lifting the waters' were things to be avoided. Yet, as Rennie's splendid stone aqueduct carrying the same canal over the Avon at Limpley Stoke reveals, our engineers were occasionally able to free themselves from this financial strait-jacket and build in the grand idiom of the Renaissance which the early French engineers understood so well.

Throughout the age of railway building in England, our engineers became increasingly subservient to commercial considerations, so that only an engineer of exceptional personality and strength of purpose such as I. K. Brunel could succeed in overriding them. It is no coincidence that Brunel was of French ancestry. The Canal du Midi revealed to me the source of the tradition that Brunel inherited and gave me a new understanding of his working philosophy. It was akin to that of Thomas Telford who once wrote: 'I hold that the aim and end of all ought not to be a mere bag of money but something far higher and far better.' Such fresh insights were a wholly unexpected bonus that I drew from my visits to the Canal du Midi.

The archives of the canal in the offices at Toulouse are singularly complete. Anyone with unlimited time and sufficient command of the French language could produce from these unique original sources a fully documented and definitive history of what is undoubtedly one of the most remarkable feats of early civil engineering in the world. Having neither the time nor the scholarship for such a major task, this book has been compiled from personal observation and inquiry, plus the sources listed in the bibliography, with the object of providing a readable introduction to the subject which I hope may stimulate more detailed study. Even so, I have found it a novel and difficult exercise which has only been made possible by many other people whose generous help is acknowledged elsewhere. The metrology of the subject posed a particular problem as some records of the canal quote pre-metric and others metric figures depending on whether they date from before or after the French Revolution. I decided to convert both to the English system with the exception of the French *tonne* (1000 kg) which was so near to the English ton as not to justify conversion. Although we are about to adopt the metric system, I did this because I believe that the English system alone is truly meaningful to the majority of English readers.

L.T.C.R.

[ONE]

The Historical Background

The completion of the Canal du Midi marked a major step forward in the art of civil engineering as applied to canal construction. But it was not the first step, so in order to appreciate what its engineers achieved and to put their work in its historical perspective it is necessary to have some idea of what had gone before.

That a gang of men or a single horse could draw a much heavier load in a boat on water than they could in a wheeled carriage on land became evident at a very early date, particularly so in times when there were no metalled roads. But water transport was at first confined to rivers which were naturally suitable, either because they were deep and slow-flowing or because, as was the case on our river Severn, powerful tides could be used to carry traffic far inland. More often than not, however, the natural gradient of a river was too steep for navigation; traffic encountered rapids and shallows which it could not surmount. This difficulty was overcome by the construction of a device variously known as a flash lock, a staunch, a navigation weir or a watergate. A weir or dam was built across the river in order to deepen the water over the shallows above. To enable boats to pass this obstruction a gap was left in the masonry of the weir which could be closed either by a lifting (guillotine) gate, a swinging gate or by a series of removable wooden boards called 'paddles' resting against a swinging beam.

For centuries, the construction of a series of flash locks of this kind was the only known means of making a fast-flowing river suitable for navigation. It was an extremely slow method, entailing very long delays while boats either waited until the weir gates could be opened or, having passed through and closed the gates behind them, for the water level to rise sufficiently to enable them to proceed on their way upstream. What put an end to this tedious proceeding and really made river, and later canal, navigation practicable was the lock, or 'pound lock' as it was at first called to distinguish it from the earlier flash lock.

The lock has often been called one of man's key inventions, but in fact it was not so much a question of invention as of evolution. Its first appearances in different countries may not have been due to the spread of a single invention but to a simultaneous process of development. For it was noticeable on any navigable river that where two flash locks were built closer together than was usual, passage through them became much quicker because the short reach of river held up or 'impounded' between them rapidly filled and emptied and so acted as an equalising pound. Thus the first pound locks were created 'by accident', as it were, and some of the first to be purpose-built in the Low Countries and elsewhere closely resembled them. Their chambers, whether they were enclosed by sloping turf banks or by masonry walls, consisted of sizeable circular or irregularly shaped basins looking like small enclosed docks. They were intended to accommodate a number of craft which passed through together and some of the earliest to be built were indeed small wet docks since they afforded communication between inland and tidal waters. They were provided with gates of the guillotine type as used on flash locks in Holland and in East Anglia, where they were introduced by Dutch engineers and locally known as staunches.

Gradually, the idea of a lock as a kind of enclosed dock whose water level could be raised or lowered at will gave place to that of a smaller structure with a chamber of masonry, usually rectangular, whose dimensions were determined by those of the craft for which it was intended. The reason for this was partly to economise in the amount of water used in lockage by equating it more nearly to the tonnage passing, and partly to avoid delays and so to ensure a more regular flow of traffic.

By the use of pound locks in conjunction with artificial navigable 'cuts' to bypass tortuous and difficult reaches, many of Europe's rivers were rendered navigable for the first time. From this type of river navigation it was a short step to the so-called lateral canal which was, in effect, one long 'cut', drawing water from a river near its headwaters but pursuing an independent course down the valley beside it and only joining it at a point where it became readily navigable, which was often within tidal limits.

On both river navigations and lateral canals, water supply presented no serious problem except in very dry seasons. It was not until the first summit level canals were built that considerations of water supply and economy in its use began to loom large. For a summit level canal is so called because it crosses the watershed (the summit) between two river valleys. This means that all the water used in lockage must be drawn from that summit level. To ensure an unfailing water supply of the necessary volume at an altitude where such a supply was not naturally available was the biggest problem that the engineers of such canals had to solve.

Commercial success depended on a satisfactory solution because the amount of traffic a summit level canal can pass depends on the amount of water available for lockage. This fact also forced engineers to design such canals with an eye to the utmost economy in the use of water. One effect of this was that the use of locks with rectangular masonry chambers of dimensions tailored to suit particular types of barge soon became universal.

In most countries optimistic proposals for summit level canals often date back a century or more before actual construction was carried out, as was true in the case of the Canal du Midi. This was because they were advanced at a time when river navigations were the rule and the engineering techniques of constructing a purely artificial waterway through difficult country had not been mastered and so posed unknown problems. Hence the natural tendency of promoters to keep the unknown to a minimum by proposing the shortest possible summit 'cut' across a watershed to link the headwaters of two rivers. Usually such proposals were not based on detailed surveys but simply on what appeared to be the shortest route on a physical map. Because they were obviously impracticable, nothing was done. But as time passed and as engineers became more experienced, so the advantages of still-water canals over fast-flowing rivers became the more apparent. Early schemes were revived in the form of more ambitious but practical proposals for much longer canals which either joined the two rivers in their lower reaches or aimed for tidal waters and so avoided the hazards of flood and drought altogether.

To illustrate this course of evolution by practical examples, the oldest pound lock in Europe is said to have been built in Holland at Vreeswijk, where the canal from Utrecht enters the river Lek, in 1373. However, in the Introduction to his *Histoire du Canal du Midi*, Général Andreossy alleges that a pound lock at Spaarndam, also in Holland, was built as early as 1285.

The Steeknitz Canal, built in 1391–98 to link the river Elbe at Lanenburg with the river Trave at Lubeck is claimed to be the oldest summit level canal in Europe. This distinction is more academic than real, however, for the nature of the terrain was such that only two locks were needed. Like the lock at Vreeswijk, these were of the basin type and were equipped with guillotine gates. They could accommodate a number of small craft and were opened twice a week. The earliest recorded pound locks in England, those built by John Trew on the Exeter Canal between 1564 and 1566, were of the same type. This last was an early example of a short lateral canal, built to bypass the difficult tidal estuary of the Exe.

Because their country was the cradle of the European renaissance, it is fitting that Italian engineers should have been the masters of canal construction in the fifteenth century. The Martesana Canal built under the direction of Bertola da Novate (1410–75) between 1462 and 1470 is a good example. Drawing its water supply from the river Adda at Trezzo, it followed the course of that river for five miles before turning westwards across the level plain of Lombardy to Milan, a total distance of twenty-four miles. It includes two locks and a masonry aqueduct of three 60 ft spans carrying the canal over the Molgora river. It would appear that this was the first canal aqueduct in the world to be designed specifically for navigation. Smaller streams are carried beneath the Martesana Canal by culverts of a design which was later to be closely followed by the engineers of the Canal du Midi.

Novate was also responsible for the Bereguardo Canal which was cut from Abbiate on the Naviglio Grande to the village of Bereguardo where there was a convenient portage for merchandise to the river Ticino. This was the first canal to be built with a considerable flight of locks – eighteen in twelve miles having a total fall of 80 ft.

It is certain that Italy takes the credit for introducing double, swinging mitre lock gates of the type still used almost universally today, but precisely who was responsible for first introducing them is open to question. The invention has been credited to the engineer Philippe Marie Visconti in 1440, while General Andreossy states that the brothers Denis and Pierre Domenico, engineers to the Venetian Republic, used double timber mitre gates for a lock at Stra, on a canal linking Padua with the river Brenta, in 1481. However, the consensus of modern opinion favours Leonardo da Vinci (1452–1519) as their inventor and he was certainly responsible for the earliest known drawing of a masonry lock with gates of this type. He is said to have introduced them on the Naviglio Interno at Milan following his appointment as engineer to the Duke of Milan in 1482.

If the older type of guillotine gate was required to close a lock chamber of any considerable breadth or depth it had to be of massive size and weight. Moreover, to provide sufficient navigable headroom it had to be raised in an overhead framework to a height of 8 ft or more above water level, and this was a slow and arduous process. Its only advantage was that it was a gate and regulating sluice combined. Swinging mitre gates, on the other hand, required some additional water-controlling device and so the lock sluice, or 'paddle' as it is commonly called in England, was provided. Although different types of sluice are to be found on canals, the oldest and still the commonest form consists simply of a miniature guillotine gate which slides in a frame and can be drawn up by a mechanism above, usually either a rack and pinion or a screw, thus uncovering an aperture behind it. These sluices are commonly fitted in the gates themselves, but to avoid deluging boats in a deep lock when it was filled, ground sluices were sometimes provided, usually above the upper gates only. In such cases the operating mechanism is mounted on the wing walls of the lock and the sluice itself admits water to an underground culvert linking the canal with the bottom of the lock chamber. Most early locks with mitre gates used gate sluices only, though Jean de Locquenghien introduced both upper and lower ground paddles on the locks of the Brussels canal as early as 1552. This pioneer example was not widely followed, however, until canal construction became much more advanced.

With its sluices and sluice mechanisms, its balance beam and 'heel post' revolving in a hollow quoin formed in the masonry of the lock wall, not to mention the Vee-shaped underwater sill that had to be built to form a seating for each pair of gates, the mitre gate was a more complex construction than the old guillotine gate, apart from the fact that it required no overhead frame. But it was far less cumbersome, being quicker and easier to operate. Consequently, as locks began to multiply and to grow deeper the advantage of mitre gates was decisive and their use spread rapidly through Europe. They appeared in France on the rivers Yevre, Cher and Auron, near Bourges, where locks having rectangular chambers measuring 90 ft by 13 ft were equipped with gates of this type in 1550, and by 1567 a triple set of mitre gates appeared at Spaarndam in Holland. Between 1571 and 1574 the first mitre gate lock in England was built at Waltham Abbey on the River Lea.

Between 1515 and 1522 the city of Milan came under the rule of King Francis I of France. That monarch was obviously impressed with the work of the Italian canal engineers, for he was himself responsible for an ambitious scheme for a new lateral canal beside the Adda river from a junction with the Martesana Canal at Trezzo with the object of providing through navigation as far as Lake Como.

When Francis I returned to France in 1516 he was accompanied by Leonardo da Vinci, and together they discussed the possibility of linking certain French rivers by canals on the Italian model with a view to opening up an inland communication between the western and eastern parts of the country. Two main proposals were considered: first, to link the rivers Garonne and Aude in the province of Languedoc, and secondly to join the rivers Loire and Saone (the Charolais or Canal du Centre). Attractive though it was, the second scheme was abandoned as too difficult and attention was concentrated on the first proposal.

One has only to glance at a map of southern France to see why the idea of linking the Garonne and the Aude, the one flowing west to the Atlantic and the other east into the Mediterranean, should have appeared so attractive even at this early date. For with their tributary streams, the Lers[1] and the Fresquel, which have their source only a few miles apart, these rivers have formed a natural corridor between the Massif Central to the north and the foothills of the

1. The modern spelling of this name is l'Hers, but I have used the older rendering throughout this book for simplicity's sake.

Pyrenees to the south to link the two seas. The existence of this natural corridor has contributed not a little to the turbulent and bloody history of the States of Languedoc, for ever since Roman times it had been a strategic route for successive armies led by such men as the infamous Simon de Montfort, the Black Prince and the Duke of Wellington. It was de Montfort who led the so-called Christian Crusade which exterminated the heretic Cathars by fire and sword with unbelievable brutality, and it was near the valley watershed at Naurouze that, in the nineteenth century, the Iron Duke signed an armistice with Marshal Soult. It has also formed an important trade route since time immemorial, and still does so as the traffic on the trunk road and railway line that now run through the valley to link Bordeaux and Toulouse with Montpellier and Marseilles bears witness.

To successive French kings and their ministers the valley appeared an obvious route for a waterway which would unite the two seas and over the years, acting on their instructions, engineers carried out surveys and came up with different proposals. Nicholas Bachelier (1485–1572) was the first of these. On the orders of Francis I he made what is believed to have been the first survey, and proposed a canal from the Garonne at Toulouse to the Aude at Carcassonne, a plan more practical but at the same time more ambitious than those of his immediate successors. Twenty years later the Italian-born Adam de Craponne[1] investigated the route,

1. De Craponne (1526–76) was the engineer responsible for the first considerable artificial waterway ever to be built in France. As this was intended for irrigation purposes only, it was not mentioned earlier in this chapter. Known as the Craponne Canal, it was designed to convey the waters of the river Durance into the dry but fertile lands of Provence and, with its branches, had a total length of 100 miles.

but then the Wars of Religion effectually halted further
progress until 1598, when Humphrey Bradley reviewed
both the original proposals of Francis I on the orders of
Henri IV and his minister the Duc de Sully. Despite his
English name, Humphrey Bradley was a Dutchman from
Berg-op-Zoom, who boasted the title of Royal Dykemaster.
His proposals for the Languedoc Canal appear to have been
the opposite of his forerunner, Bachelier – less ambitious
but at the same time less practicable. For example, he is said
to have proposed canalising the little river Lers west of the
summit at Naurouze which would have involved major
works along the floor of a valley which was then notoriously
boggy. It was correctly judged to be impracticable. But the
major problem which so far remained unsolved was this:
no matter how far these schemers tried to reduce the
magnitude of such a project by proposing to make the
tributary rivers Lers and Fresquel navigable, they could not
escape the fact that to link the two would mean carrying a
canal over a 620 ft summit level at Naurouze.

It was the question of how to provide this short summit
with an adequate and reliable supply of water that defeated
them. This problem was aggravated by the semi-tropical
climate of Languedoc. Long periods of drought and extreme
heat in the height of summer would be punctuated by
infrequent but violent storms. It was these climatic extremes
which ruled out the idea of economising in construction cost
by making the local rivers navigable. East of the summit,
summer droughts would shrink the Fresquel, the Aude, the
Orb and the Hérault to shallow, unnavigable streams
meandering between wide gravel spits. Yet, in a matter of
hours, a storm over the Massif Central could transform these
rivers into raging torrents. It was obvious that such

treacherous rivers were best avoided by prolonging the
length of canal along higher ground clear of their flood-
prone valleys. But such a policy magnified the problem of
water supply and also raised another one, which was how to
cross these rivers. For with the exception of the Aude,
which rises in the Pyrenees, all have their source in the
Massif Central and drain into the Mediterranean, thus
intersecting the line of route.

Such were the formidable problems that led Sully to
counsel the king that the Languedoc Canal scheme should
be shelved and that attention should be devoted to a more
modest proposal further north where the results of
Humphrey Bradley's investigations appeared to be more
promising. This proposal was for a canal to link the Loire
with the Seine. The scheme was not only attractive commercially, but it might ultimately form a branch of that second
Atlantic-Mediterranean route that Francis I and Leonardo
had first envisaged. Sully's decision was a wise one, for
although the Canal de Briare which ultimately resulted
from it was only 21¼ miles long it was to serve as a working
model for those who would ultimately engineer the Canal
du Midi.

The plans originally drawn up in 1603 for this Loire –
Seine link were typical of their period in that they involved
extending river navigation as far as possible on each side of
the watershed, thus reducing the length of canal to be cut to
a minimum. From its junction with the Loire at Briare, the
tributary Trezée was to be made navigable for 10 miles to
the village of Breteau. Similarly, on the other side of the
summit plateau, the Seine tributary, the Loing, was to be
made navigable for 26 miles from Rogny down to Montargis. The summit level was 266 ft above sea level and it was

thought that a total of forty-eight locks would be required.

This scheme was put out to tender and in February 1604 the contract was awarded to Hugues Cosnier. Cosnier was more than a contractor; he was an engineer of genius and remarkable tenacity of purpose. Having examined the line with care he decided that the plan he was to execute was defective in two most important respects: it made no provision either for water supply to the summit level or for the protection of the works on the rivers Trezée and Loing against flooding. Traffic on such river navigations is always subject to interruption by flood or drought and it was for this reason that James Brindley pronounced that the only useful function of rivers was to supply canals with water. Nearly 200 years earlier Hugues Cosnier obviously reached the same conclusion for he produced an alternative scheme for a new canal all the way from Briare to Montargis, cutting out the proposed river sections altogether. He also proposed a realignment of the summit level which obviated the deep cutting previously envisaged and at the same time made better provision for its water supply. This new summit level was nearly three miles shorter than the original scheme.

The water supply arrangement planned by Cosnier consisted of a feeder channel $3\frac{1}{4}$ miles long from a headwater intake near the source of the Trezée to a small lake called the Étang de la Gazonne which acted as a reservoir. From this lake, the water could fall directly into the summit level of the canal. To ensure an additional reserve, Cosnier made an ingenious provision so that part of the summit level itself could act as a reservoir. A length of $1\frac{3}{4}$ miles at one end of it was given an additional depth of from 8 to 9 ft, this deepened section being divided from the rest of the summit by a stop lock. The depth of the gate sills at each end of this section was sufficient to allow up to 5 ft of water to be drawn from it without interfering with navigation. It will be recalled that Robert Whitworth adopted the similar expedient of deepening the whole of the short summit level of the Leeds & Liverpool canal, though whether he was consciously following Cosnier's example we do not know.

Cosnier also originated another principle which was to be widely adopted by subsequent generations of canal engineers. This was to concentrate the locks as far as possible so as to allow the maximum length of level pounds between them. Such an arrangement not only simplifies lock supervision and maintenance, but it also improves traffic working and water regulation.

This revised scheme of Cosnier's was accepted by the Royal Council in December 1604, and 6000 troops were drafted to the area to provide the labour force. Work then proceeded apace until 1611, when the Duc de Sully was forced to resign following the assassination of the King in the previous year and a new regime appointed a commission, which included Humphrey Bradley, to report on the state of the works and the cost of completion. Although they reported enthusiastically on a canal which they found more than three-quarters complete and recommended that the work should go on; although Cosnier himself offered to carry on at his own risk in return for the right to collect tolls for the first six years of operation, owing to wars and weak government nothing further was done, and the works lay derelict for seventeen years. Though he found engineering employment elsewhere, it must have been a bitter blow to Cosnier who, at the time the work was stopped, had built thirty-five out of the total of forty locks.

In accordance with Cosnier's principle of concentrating

the locks, many of these were double locks or 'risers' as
they are called in England; that is to say they consisted of
two successive lifting chambers divided by an intermediate
sill and gates. There were also three multiple 'staircases' of
locks on the same pattern, that at Rogny where the canal
fell steeply into the Loing valley having six chambers with a
total fall of 65 ft – certainly an impressive engineering work
and one quite unparalleled at this date. All the locks were
built with masonry side-walls 6 ft thick, and their levels
were controlled entirely by ground sluices and culverts
after the model of those introduced by Locquenghien on the
Brussels Canal which Cosnier had visited.

In 1628 two engineers, Francini and Le Mercier, re-
examined the canal and strongly recommended completion,
but with the proviso that an additional water supply to the
summit level should be provided, the reason being that
large lock staircases such as that at Rogny consume a
prodigious amount of water, particularly when, as in this
case, their chambers were large enough to admit craft
105 ft long and 15 ft beam. They proposed a second feeder
channel from the headwaters of the Loing at St-Privé. This,
like the first feeder, would flow into a small lake situated
beside the summit pound but at a higher level, so that it
could act as a reservoir. Cosnier offered to construct this
second feeder but did not live to do so. The old engineer
died in 1639 and his master-work lay derelict for another
eight years until, a year before his death, Guillaume
Boutheroue and Jacques Guyon obtained letters patent from
Louis XIII authorising them to complete the canal and to
pay off outstanding debts in return for ownership. For this
purpose they formed a company and, under the direction of
Guillaume's brother, François Boutheroue, the remaining

1   Pierre Paul Riquet, from a portrait in the Musée Paul Dupuy at
Toulouse

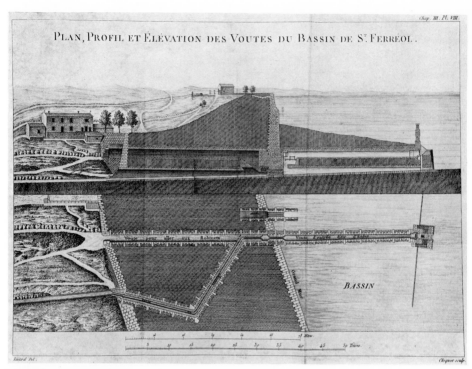

PLAN, PROFIL ET ÉLÉVATION DES VOUTES DU BASSIN DE St FERRÉOL.

Voute pour aller aux Robinets   Voute contre d'ufice d'Eaue

Grand Mur

BASSIN

Lecleré Del.   *Cloquet sculp.*

2, 3   The St Ferreol Dam: *top*, cross-section and plan;
below, looking along the crest of the dam, spillway and upper
sluice in foreground

five locks were built, the original works repaired and the canal completed and opened in 1642, precisely as designed by Cosnier thirty-eight years before. As Francini and Le Mercier had forecast, however, the one flaw in his design was that the water supply to the summit proved inadequate and, as they had recommended, a second feeder channel, 13 miles long was dug from St-Privé and completed in 1646. It was a remarkable feat of early surveying since its total fall is only $5\frac{1}{2}$ ft or 5 inches to the mile.

The Canal de Briare was an immediate commercial success, earning 13 per cent on its capital and carrying an annual average of 200,000 tons of traffic for the next 150 years. Moreover, unlike the sad history of our English canals, the annual tonnage carried doubled during the nineteenth century and trebled during the first half of this century. Yet in its early days there remained one serious snag. For all his pioneering vision Cosnier had not been quite bold enough. For the navigation of the Loing between Montargis and the Seine was obstructed by no less than twenty-six flash locks constructed beside the mills on that river. Here, as elsewhere, the mill-owners objected to the replacement of these by pound locks, arguing, not without reason, that they would interfere with the working of their mills by causing the water to back-up in their tail-races. Precisely the same situation occurred in this country where a new canal joined an archaic river navigation,[1] but in this case the intolerable delays so caused were soon remedied, for by 1723 a new lateral canal had been built from Montargis to the Seine.

Compared with the 149 miles of the Canal du Midi, the

1. The case of the Thames & Severn Canal and the Upper Thames Navigation, for example.

Canal de Briare seems a very minor undertaking. Yet it was the first considerable summit level canal in the world and in its day it represented the most advanced development of the art of canal construction. Its success undoubtedly encouraged those engineers and promoters who still dreamed of a water link between the Garonne and the Mediterranean. Although the times were unpropitious for canal construction, as the long delay in the completion of the Canal de Briare bears witness, that dream did not die. In 1633 an engineer named Etienne Tichot, who is described by La Lande as 'engineer du Roi', presented to Cardinal Richelieu a plan for a canal from the Garonne at Toulouse to the Aude near Narbonne. A little later another scheme appeared, this time joining the Aude at Trèbes, a village on the Aude a little to the east of Carcassonne.

Even on the shorter westerly course of the canal between the summit at Naurouze and the junction with the Garonne opinions were divided. Although the city of Toulouse may have seemed an obvious western terminus, there were those who advocated making the little river Lers navigable throughout which would have meant by-passing the city altogether; for the Lers skirts the eastern boundary of Toulouse before flowing into the Garonne near the small town of Grenade, a few miles downstream from the city.

Finally another proposal, which would have given Toulouse an even wider berth, must be mentioned here because it is not without significance in the light of subsequent events. It grew out of a scheme launched by a certain Pierre Borel, Regent of the College of Castres, for providing Castres with a water route to the Garonne by making the river Tarn and its tributary the Agoût navigable. By the construction of thirty-one locks, the Tarn was eventually made navigable for 91 miles from its junction with the Garonne at Moissac to a point 5½ miles upstream from Albi. Indeed, this Tarn Navigation survived in use until 1888 after which it subsequently succumbed to the effects of flood and drought.

When work began on the Agoût Navigation, from the junction of that river with the Tarn at St Sulpice to Castres, the suggestion was made that it might form the western section of a waterway between the two seas. This was to be achieved by making the river Sor navigable from its junction with the Agoût at Vielmur-sur-Agoût, a little below Castres, to a point near the little town of Revel, whence a canal would be cut to Naurouze.

About this time, the unfortunate Pierre Borel, who was a Protestant, lost his position at the College of Castres when that institution was taken over by the Jesuits and nothing more is heard of him. Whether he was responsible for this further proposal we do not know although there is a strong supposition that he was. If this was the case, he evidently had an intimate knowledge of the local topography and came within an ace of solving the key problem that had bedevilled all plans for a canal between the two seas for so long.

The river Sor rises in the Montagne Noire, from which it flows in a north-westerly direction to join the Agoût. This Montagne Noire is a mass of granite, rising to a height of 4000 ft, which forms the extreme south-westerly promontory of the Massif Central. Compared with the parched lands to the south-east, rainfall over the mountain is heavy and the many streams that fall from it have scored deep valleys in its thickly wooded slopes. Some of these streams flow southwards to swell the river Fresquel and so ultimately find their way to the Mediterranean; others, like the Sor, take an opposite course, flowing north and west towards the Atlantic. From the foot of the mountain a limestone plateau known as the Lauragais extends southwards to the foothills of the Pyrenees. Here the rivers Lers and Fresquel have their sources, so that it is this plateau which forms the watershed between the Atlantic and the Mediterranean. Its lowest point is the Col de Naurouze, sometimes called the Seuil (threshold) de Naurouze. Borel, if indeed it was he, had grasped the fact that, if the Sor were to be made navigable, it would be perfectly possible to link that river with Naurouze by a circuitous canal following the contours of the Lauragais round the headwaters of the Fresquel. But this would make an extremely roundabout route for a waterway between the two seas and, although it might bring benefit to the cities of Castres and Albi, it would by-pass the capital of the province, Toulouse, altogether.

In 1662 a party of local government officials from Castres inspected the state of the new navigation between Montauban and Castres. They found that work on the Agoût had run into the kind of difficulties that all too commonly beset such works at this period. Local landowners were strenuously resisting the cutting down of trees and other works necessary in order to straighten the course of the river. Local mill-owners were obdurate, claiming that the navigation would interfere with the working of their mills. One of the latter, a M. de Garribal, was particularly obstructive, refusing to permit the working of a lock which had been newly built beside his mill. The inspection party found that things had reached such a total impasse that an appeal was made to the king's chief minister, Colbert, by the three dioceses of Albi, Castres and Lavaur begging him to use his great influence to ensure the successful completion of the works.

By now, the projected canal between the two seas had been talked about and reported upon in the States of Languedoc for a hundred years without a single sod being turned. The people in the cities and towns along its proposed route, Toulouse, Castelnaudary, Carcassonne and Narbonne, had seen so many engineers and surveyors come and go without result that it must have seemed to them that the canal was no more than an impossibly grandiose dream. Yet the dream was about to become a reality. This was due to two factors: an unusually favourable political climate in France and a man on the spot of quite exceptional vision, imagination and drive to get so formidable an engineering project off the ground. It was 'le Roi Soleil', Louis XIV, and his great minister, Colbert, who between them provided the right climate, while the man on the spot was Pierre Paul Riquet, ably assisted by Pierre Campmas, the *fontainier* of Revel, and a young engineer named François Andreossy.

[TWO]

Conception

Pierre Paul Riquet was born at Béziers on 29 June 1604. His family is said to have been Italian in origin and, under the name of Arrighetti, to have moved to southern France in the service of the Gibelins in 1268. The family multiplied and settled in Provence and Languedoc, changing their name to Riquetti and finally to Riquet. In the sixteenth century, Regnier Riquet founded a branch of the family who became Comtes de Caraman in Languedoc. Like so many immigrants, the Riquets soon identified themselves completely with their adopted country. So much so that the youthful Pierre Paul, to the dismay of his teachers, refused to learn Greek, Latin or even French, preferring to speak in the local tongue, Occitan, or 'la Langue d'Oc', which gave the Province its name.

Pierre Paul's grandfather, Nicholas, was a master-tailor who had improved the family fortunes by marrying Béatrice Bordier, a girl of wealthy parents. His father, Guillaume, practised as a lawyer in Béziers and succeeded in amassing a fortune by means that were not too scrupulous. He also took advantage of local political disorders to get himself appointed one of the thirty 'managers' or counsellors of the city and so wielded considerable influence in Béziers. In January 1618 one of the many schemes for a canal between the two seas came up for consideration by this council and Guillaume Riquet was one of the majority who voted against it, perhaps because its proposed route to the Mediterranean led through Narbonne and not through Béziers. But this – and doubtless other schemes for a canal to link the two seas – were discussed in the Riquet household and the idea fired the imagination of the youthful Pierre Paul.

Riquet was educated at the Jesuit College in Béziers where, although his refusal, or inability, to learn languages made him unpopular with his teachers, he displayed a great interest in, and a natural flair for, the sciences and mathematics. This was recognised and encouraged by his godparent, Father Portugniares, but the emphasis seems to have been on the mathematics for both his father and godfather saw in the young Riquet a future financier and man of business. As Riquet admitted later, he received no formal engineering training.

Like his grandfather before him, Riquet married money. When he was only nineteen years of age, he was married to Catherine de Milhau, daughter of a wealthy bourgeois family in Béziers whose dowry was such that he was able to purchase the château and estate of Bonrepos, near the little village of Verfeil twelve miles to the east of Toulouse on the slopes of the valley of the river Girou. Seven years later, in 1630, through the influence of Father Portugniares, Riquet was appointed a collector of the salt tax in Languedoc, and it was not long before he became Lessee General (*Fermier Général*) of this tax for the entire province. The position was a lucrative but onerous one which entailed a great deal of travelling throughout Languedoc, and in this way Riquet gained an intimate knowledge of the country through which the Canal du Midi would eventually pass. In 1632 he appointed as his deputy an ex-school-friend of his named Paul Mas, who had become a Doctor of Law and was a brilliant lawyer in Béziers. In the same year Riquet's father died and he inherited the principal part of his estate, making the family house in the Place St Felix in Béziers the centre for his business in south-eastern Languedoc.

A tax on salt had been first introduced in France in 1206 and had become one of the chief sources of state revenue. Riquet's office was an extremely profitable one and, aided by the efficient Paul Mas, who became his brother-in-law, he made so much money that he was able to embark on the business of a military contractor, supplying the King's armies in Cerdagne and Roussillon. This contracting business was also very profitable, with the result that by the time he was fifty Riquet had amassed a fortune of several million livres. In 1665 he acquired a town house in the Place St Pantaléon in Toulouse, but he liked best his country estate of Bonrepos; there only did he feel truly at home, and as the years went by he was able to spend an increasing amount of his time there.

Most men, having achieved such material success, and with so able a deputy, would, at the age of fifty, have been content to retire to Bonrepos to lead the life of a country squire. But Riquet was an exceptional man. The quiet life was not for him, and the work that would make him famous was still before him. Few men can have embarked, with all the burning enthusiasm and restless energy of youth, upon so great a task so late in life. The dream of a canal to link the two seas which had fired his imagination as a boy in Béziers had never left him; the many abortive proposals that had come and gone had served to keep the idea alive for him and now he had the wealth and the leisure to consider the idea seriously and to ponder whether the dream might become a reality. He had visited the Canal de Briare and, having admired Cosnier's great lock staircase at Rogny and his other works, there was no longer any doubt in his mind that the construction of a canal from sea to sea was practicable; the great problem was how to supply it with water. This was the rock on which all previous schemes had foundered.

It is important to remember that the Canal du Midi, like Telford's later Caledonian and Gota canals, was primarily intended to obviate a long sea passage and therefore had to be designed to pass the seagoing trading craft of the day. Admittedly in Riquet's time such craft were small but, nevertheless, this meant that, compared with the major part of the English canal system which was designed solely for inland transport, the Canal du Midi had to be built on a grand scale. A tremendous volume of water would necessarily be consumed in lockage. In addition, the amount of water lost by evaporation would be far greater than in the more temperate climate of England, although we do not know whether Riquet took this second factor into account. But in any case, he realised that a copious supply of water must be fed into the proposed summit level at Naurouze, a supply which must not fail during the hot, dry summers.

On first consideration, Riquet accepted the proposal that the canal should fall eastwards from the summit to join the Fresquel between Castelnaudary and Carcassonne. From this junction the rivers Fresquel and Aude would be made navigable to a point near Sallèles where the Canal de la Robine joined the Aude. This ancient canal (it was reputedly dug by the Romans) runs through the heart of the ancient Roman city of Narbonne to enter the Mediterranean at Port la Nouvelle.

For the western section of the canal between the summit and the Garonne, Riquet carefully weighed the merits of three proposals which had been made. La Lande enumerates these as follows:

1. A canal from the summit northwards to the Sor and thence via the rivers Agoût and Tarn to the Garonne at Moissac.

2. A short canal from the summit to the river Lers which would be made navigable to its junction with the Garonne at Grenade.

3. A longer canal from the summit to join the Garonne at Toulouse.

Of these alternatives, Riquet chose the third. It was to him unthinkable that such a great work should by-pass the capital of the province of Languedoc, while he rightly decided that to make use of the Tarn and its tributaries would mean that navigation would become liable to interruption by flood or by drought. They were best avoided. Nevertheless, it was the first scheme that gave him his inspiration, promising a possible solution to a water supply problem that all his life had seemed insuperable. For if a branch canal could indeed be built from the river Sor near Revel to Naurouze, might it not be used to convey the waters of the Sor into the summit level? This was something that had never occurred to the original authors of the scheme. So concerned were they with the fortunes of Castres and with the canalisation of the rivers Sor, Agoût and Tarn to the north of Revel that the idea of directing the waters of the Sor due south simply had not crossed their minds. But it made sense to Riquet.

The Sor at Revel is a very small river, that place being not far from the river's source in the Montagne Noire. Moreover, any proposal to divert a major part of its flow southwards would be bound to rouse fierce opposition from the owners of mills situated further downstream. They would demand compensation water. This being so, Riquet came to the reluctant conclusion that the river Sor alone could not be relied upon to provide a dependable all-the-year-round supply – it must be augmented. Riquet reasoned that there was only one possible source from which such extra water might be drawn and so he turned his attention to the steep wooded slopes and deep valleys of the Montagne Noire.

By 1662 Riquet had become convinced that he had the right answer, but if he was to convince others there was no time to be lost, for by this date the Sor–Agoût–Tarn route had the combined backing of the dioceses of Albi, Castres and Lavaur, and Mgr d'Anglure de Boulemont, Bishop of Castres, had solicited Colbert's support for the scheme. Revel and the Montagne Noire were within easy riding distance of Bonrepos, so Riquet rode over to consult the men best calculated to advise him – Pierre Campmas, the *fontainier* of Revel, and his son, Pierre junior, who was as knowledgeable as his father.

There is no precise English equivalent for the occupation of *fontainier*. Pierre was responsible for the maintenance of the water supplies of Revel and district. He was also responsible for regulating and maintaining the local water courses in order to minimise flooding and so avoid damage to property. He therefore knew the valleys and streams of the Montagne Noire like the back of his hand. Amongst the villagers of Verfeil, Riquet was a popular figure, for he was a straightforward, humorous and unaffected man who always got on well with country characters like Pierre. Moroever, his enthusiasm was infectious, so that it did not take the *fontainier* long to grasp the importance of Riquet's idea and to

become as enthusiastic about it as he was. The two men took
to each other at once and this was the first of a number of
meetings in the course of which, like two prospectors, they
quartered the Montagne Noire, Riquet on horseback, Pierre
striding out with a staff in his hand. After these expeditions
Riquet would often spend the night with Pierre in his
cottage at Revel.

From the southern slopes of the Montagne Noire a series
of streams, the Alzau, Vernassone, Lampy, Lampillon and
Rieutort fall to join the waters of the eastward-flowing
river Fresquel. As a result of their explorations, Riquet and
Pierre concluded that the waters of these streams could be
tapped by a feeder channel. This would extend along the
contours from the easternmost Alzau intake at a height of
2120 ft above sea level to a place called Conquet, where the
feeder would be situated on the crest of the east–west
watershed and could conveniently discharge its waters
down a steep slope to swell the infant river Sor. This would,
they judged, make the supply to the feeder canal more
reliable and at the same time furnish enough compensation
water to keep the local mill-owners happy. But Riquet was
still not entirely satisfied that this expedient alone would
ensure a reliable supply to his proposed canal during the
dry months of summer. The deep and narrow valleys of the
Montagne Noire naturally lent themselves to damming and
he conceived the idea of a series of reservoirs, linked by
natural channels to his mountain feeder. In this way the
rainfall over the mountain could be stored up and then
released at will during the summer. It is clear from subsequent correspondence that he was thinking along these lines
at this time although, in the event, he would be responsible
for building only one great dam.

It was apparently at this juncture that the third member of the team, François Andreossy, first appeared on the stage. He was recommended to Riquet as a likely young man by the Bishop of Castres, d'Anglure de Boulement, a fact which suggests that Riquet had already mentioned his idea to de Boulement in a not unsuccessful attempt to talk him out of the Castres scheme. François Andreossy (1633–88) was a citizen of Narbonne but was born and educated in Paris, where he studied science and civil engineering, with a special emphasis on hydraulics. Having completed his education in the capital he returned to Narbonne, but in 1660 he left there for Italy, where he had family connections, so that he could study the canals of Lombardy and Padua. At the time with which we are concerned he had but lately returned, filled with enthusiasm for canals, from this Italian visit. Andreossy supplied the professional expertise which Riquet lacked. Unlike Pierre Campmas who, having contributed his store of local knowledge and experience to the project, plays only a minor part in the story of the Canal du Midi, Andreossy became Riquet's right-hand man throughout the whole period of construction.

Who should be entitled to the major share of the credit for the Canal du Midi, Paul Riquet or François Andreossy? This is a question that has always been disputed. By the time construction of the canal began, Riquet was over sixty years of age, and although his memory is green in Languedoc, although there are statues and columns which perpetuate his achievement and streets named after him, it strikes an outsider as inherently improbable that an elderly tax-gatherer without, on his own admission, any previous experience in the art could ever have been responsible for the greatest feat of engineering in Europe. And to the obvious question which this fact prompts, the answer seems ready to hand in the person of the young, enthusiastic and professionally trained engineer François Andreossy. It is tempting to see in him Riquet's James Brindley. This matter was long disputed between the descendants of the two men. Général Andreossy, the great-grandson of François, evidently had no doubts on this score when he wrote in his *Histoire du Canal du Midi*: 'François Andreossy, l'ingénieur de ce grand ouvrage, s'en fit enlever la gloire par l'entrepreneur Paul Riquet; et le talent, moins heureux que le crédit, fut privé de toutes les récompenses.'

The account of the inception of the canal project given by Général Andreossy in his book differs completely from that written in 1778 by La Lande, which was approved by the Comte de Caraman and the Baron de Bonrepos, the descendants of Riquet and the proprietors of the canal at that time. According to the Général it was Andreossy who first conceived the idea of the feeder system which solved the water supply problem, and he communicated it to the wealthy Riquet with the object of winning his support for the project.

Ancestor-worship is one of the most potent falsifiers of history. Knowing this, and faced with these two conflicting accounts, it is not easy for a stranger, and a foreigner at that, to discover the true course of events which took place three centuries ago. However, if a historian ponders a series of apparently disconnected and conflicting facts long enough it may suddenly dawn on him that, like the pieces of a jigsaw puzzle, they can only be fitted together in a certain logical way and in no other. This chapter represents such a reconstruction. It is also abundantly clear from his letters that Paul Riquet was no ordinary tax-gatherer and, indeed, no ordinary man. Nor was he merely an entrepreneur, despite his lack of engineering training. If the canal project had failed, who would have carried the blame? From all the evidence, the unbiased answer must be Riquet and therefore, because it did not fail, we cannot withhold from him the credit. On the other hand, there is no doubt that Andreossy, along with other professional engineers and surveyors, did a great deal of the spade work by translating Riquet's ideas into practice, and that without such help Riquet could not have succeeded. Perhaps the most useful analogy is to liken Riquet to the conductor of an orchestra of which Andreossy was the leader; it was Riquet who had the inspiration which made the project practicable in the first place, and it was he who supplied the tireless drive and enthusiasm, coupled with the unique organising ability that was needed to see such a vast undertaking through to a successful conclusion.

Having carried out his fruitful exploration of the Montagne Noire, Riquet's next step was to build a model canal in the grounds of Bonrepos with the assistance of Pierre Campmas and the new recruit Andreossy. The old house of Bonrepos, built of that rose-red brick so characteristic of the Toulouse district, was surrounded, or partially surrounded, by a moat which was fed from two small lakes at different levels in the park. These lakes, which may originally have been millponds, were ideally suited to Riquet's purpose and here he is said to have constructed a miniature canal complete with locks, weirs, feeder channels and even a tunnel. It is also said that remains of these works, much overgrown, can still be found in the grounds of Bonrepos. One is irresistibly reminded of the miniature lock which James Brindley built in the grounds of his house

at Turnhurst before he embarked on the construction of his
Trent & Mersey and Staffordshire & Worcestershire
canals. Precisely why Riquet or Brindley should have found
it necessary to build such trial constructions is by no means
clear. There were plenty of pound locks in England for
Brindley to study and compare, while Riquet was already
familiar with the Canal de Briare, a full-scale example of a
summit level canal which, in 1662, had been working succes-
fully and profitably for twenty years. Nevertheless, he
appears to have needed this model before he could satisfy
himself and others that his ideas were practicable.

While all this activity was going on at Bonrepos, Riquet's
friend de Boulemont, Bishop of Castres, was appointed
Archbishop of Toulouse. As such, he became the most
powerful and influential man in Languedoc, chief representa-
tive of both the spiritual and the temporal powers of God
and King. In the late autumn, the new Archbishop visited
Bonrepos, and witnessed a demonstration of the working
of the model canal by the faithful Pierre Campmas. Riquet
then explained his plans in detail and de Boulemont displayed the keenest interest, suggesting that he would like to
visit the Montagne Noire with Riquet. This was arranged,
and Riquet subsequently led a party consisting of the
Archbishop, the Bishop of Saint-Papoul and several other
local notables over the line of his proposed high-level
feeder channel. De Boulemont promised to use his influence
with Colbert to further the scheme, but suggested that, as a
first step, Riquet should himself write to Colbert explaining
that he did so on the Archbishop's instruction. This historic
letter, which was destined to set the great scheme in motion
at last, was duly written and sent, dated at Bonrepos,
26 November 1662. In free translation it read as follows:

4, 5   The St Ferreol Dam: *top*, entrance to 'Voûte d'entrée d'Enfer' in lower dam wall. The path on the left leads to the entrance to the valve chamber; *below*, the spillway and fountain below the dam

6, 7   The Rigole de la Montagne: *top*, Les Cammazes
Tunnel, built by Vauban *c.* 1685; *below*, the end of
the Rigole: 'Rigole de Ceinture' (left), supply to
St Ferreol, right

I am writing to you from this village on the subject of a canal which could be made in this Province of Languedoc for the connection of the two seas. You will be astonished that I dare to speak of such a thing of which I apparently know nothing and that a salt tax collector should be engaged on matters of surveying. But you will excuse this step when you know that it is at the order of the Archbishop of Toulouse that I write. It is some time since the said Lord did me the honour of coming to this place, either because I am a neighbour and admirer of his, or in order to learn from me the way of making this canal – because he had heard tell that I had made a particular study of it. I told him all I knew of it and promised to go to see him at Castres on my return from Perpignan and to lead him from there by ways which would enable him to see the possibility of it. I did this and the said Lord, in the company of the Bishop of Saint-Papoul and several other distinguished people, visited all the things which were found as I had told them, [whereupon] the said Lord Archbishop charged me to draw up an account of it and to send it to you. It is enclosed here, but in rather poor order because, knowing no Greek or Latin and only with difficulty knowing how to speak French, it is not possible for me to express myself without faltering.

Also, I undertake this in order to obey and not of my own volition. All the same, if you will please read my account you will judge that this canal is possible, that it is, in truth, difficult as to cost but that, having regard to the good that should come from it, one ought to weigh the expenditure less highly.

Until today, the suitable rivers to supply it had not been thought of, no easy route for this canal was found because those one had imagined easy [involved] the insuperable difficulties of reversing the flow of rivers and of machines for lifting the waters. So you will believe that such difficulties have always discouraged people and postponed the execution of the work.

But now, my Lord, that easy routes have been found, as well as rivers able easily to be diverted from their ancient beds and conducted into this new canal by natural fall away from their usual gradient, all difficulties cease excepting that of finding money to cover the cost of the work.

You have for this a thousand means, my Lord, and I suggest to you two more still in my attached memoir in order to incline you most readily to this work which you will judge very advantageous to the King and to his people when it pleases you to consider that the facility and safety of this navigation will make the Straits of Gibraltar cease to be a necessary passage; that the revenues of the King of Spain from Cadiz will be diminished by it and that those of our King we shall be augmenting by so much on the Treasury leases and from the import of merchandise into the Kingdom – apart from the tolls which will arise from the said canal which will bring in enormous sums, and his Majesty's subjects will profit from thousands of new commercial enterprises and will draw great advantage from the navigation.

If I hear that the project gives you satisfaction I will send you a financial estimate marked with the number of locks which it will be necessary to make and with exact calculations in fathoms of the said canal, either in length or in breadth.

Then follows a memoir of which the most significant passage reads:

… But what seems to me most important is to have enough water to fill it [the canal] up and to take this to the right place where the summit level is, which can easily be done, using the river Sor near the town of Revel, whence it will flow by natural gradient because there is a 54 ft difference in level between the said Revel and the watershed, the country being flat and without any hills. It is also easy to drive the brook called Lampy into the bed of the Revel river, distant about 1500 paces from one another. It is just as easy to drive into the Lampy another brook called the Alzau about five quarters of a league distant, and consequently, several other brooks which one finds along that conduit, so that, joined together, the vigorous and durable supply they constitute will make up a big river which, driven to the watershed, will supply both sides of a canal six feet in depth and 54 ft wide throughout the whole year so that navigation on this canal would be easily carried on.

This, in a nutshell, was the key to the success of Riquet's great scheme and we may imagine the anxious thought that was put into this missive and the drafting and redrafting that went on in the study at Bonrepos before Riquet was finally satisfied and sent it off. It certainly had the desired effect. The great Colbert must have sensed that the man behind the letter was no rural dilettante concocting visionary proposals by his fireside, but a likeable and forceful character who had hit upon the first really practicable proposal for a canal between the two seas. He could even be the man who, given sufficient financial backing, might possess enough enthusiasm, ability and drive to see such a project through to a successful conclusion.

Jean-Baptiste Colbert (1619–83) was the son of a Reims draper who worked his way into the Royal service under Le Tellier and Mazarin. His efforts to establish state industries throughout France included founding a cloth-weaving industry, supervised by Dutch craftsmen, at Carcassonne. He was well aware that no industry could flourish without efficient communications. He therefore ordered the repair and better maintenance of roads and bridges and under his aegis the first paved road was completed between Paris and Orleans. Another target for his reform was the complex internal toll structure of his country, and he succeeded, against great opposition, in establishing uniform toll rates throughout the central provinces. The prospect that a water route from the Atlantic to the Mediterranean might be achieved at last naturally appealed strongly to such a man.

So the result of this letter was that Riquet and de Boulement were summoned to an audience with the great man in Paris, travelling thither in the Archbishop's coach. Riquet and Colbert seem to have taken an immediate liking to each other and this interview marked the beginning of a long, happy and fruitful association in which Colbert patiently supported Riquet almost to the end, not only with money, but with kindly and encouraging letters whenever things went wrong and the mercurial Riquet became despondent.

Although Colbert favoured Riquet and his scheme, as the King's finance minister he had to proceed circumspectly. He mentioned the matter to Louis XIV who was greatly taken with the idea that his reign should be marked by the execution of a project of such scale and grandeur. On the advice of Colbert the king made an Order in Council, dated 18 January 1663, appointing a Royal Commission to investigate the soundness of Riquet's proposals. The Commissioners, in turn, appointed four surveyors to assist them, and these included the Geographer Royal, one Jean Cavalier, and François Andreossy.

With one exception, the members of this Commission were more distinguished for their eminence in public life than for any knowledge of canal engineering. The significant exception was Henri de Boutheroue de Bourgneuf, the son of that François Boutheroue who had completed Cosnier's work on the Briare Canal. Henri, who had succeeded his father as a Director of the Briare, was appointed principal expert on the Commission, a fact which must have greatly pleased Riquet.

During the next twenty months, Riquet was engaged

with Henri Boutheroue in going over the line of route of
the canal and its feeders in order to prepare a preliminary
estimate of the extent of the works and of their cost. This
was then carefully checked before presentation by the
Chevalier de Clerville, the general commissioner of forti-
fications. De Clerville apparently accepted the estimate as
the unaided work of Henri Boutheroue, an excusable
error in the light of Riquet's lack of previous canal experience. What precise part Boutheroue played in this
preliminary survey we have no means of knowing, but
whether it was due to his influence or not, it is apparent
from a letter that Riquet wrote to Colbert on 20 October
1664 that his ideas had changed considerably since he had
written his original letter two years before. On the construction of his proposed feeder system he refers to 'costly
banks' and it is clear that he now feared that it would prove
much more expensive than he had first thought to provide a
supply which 'would keep the canal full and the country
round irrigated during eight months of the year'. If this
view reflects Boutheroue's pessimism, it was later to prove
unfounded. On the question of conserving water in the
Montagne Noire for use during the summer months, Riquet
had formed a much more realistic notion of the amount
required for he speaks of 'storing water during the winter in
fifteen or sixteen reservoirs for use during the four months
of drought'.

So far as the main line of the canal was concerned, though
Narbonne was still the eastern objective, Riquet now
abandoned the old concept, which he had inherited, of
making the rivers Fresquel and Aude navigable and so
reducing the length of the canal line. 'Speaking of the river
Aude', he wrote, 'it is irregular, strewn with stones and can

only be used by light boats. Better to make use of its water and build a new canal alongside, at some distance from it and high enough not to fear its floods. The new canal will connect with the old Robine [canal] and flow into the Mediterranean.' This change of plan obviously owed a great deal to what Boutheroue must have told him about the shortcomings of the Loing Navigation between Montargis and the Seine and the adverse effect of the delays so caused on the Canal de Briare.

In November 1664 Riquet formally presented the results of his survey to the Commissioners. Included among the documents was a map showing the feeder system and the main line of navigation between Toulouse and Narbonne. This may well be the one that Général Andreossy reproduces in his book, claiming it to be the oldest known map of the Canal du Midi. It was, he says, discovered amongst some old papers after the lapse of 150 years and, although it bears no date or attribution, he alleges that it was dated 1664 and that it is the work of his great-grandfather. It is a crude and archaic affair, showing the feeder system in plan and the main canal in elevation as a series of steps. If these steps are not merely figurative but are actually intended to depict each lock site, then the map is decidedly optimistic, for they total 56, 11 ascending from Toulouse to the summit and 45 descending to Narbonne. As built, the canal would have 101 locks, 26 ascending from the Garonne and 75 on the descent to the Mediterranean.[1]

The Commissioners reported on 22 January 1665, and it

1. These figures represent the number of lock chambers. The issue is somewhat confused by the fact that in France double and staircase locks are frequently counted as if they were singles. A total of 99 lock chambers is now in use.

is clear from this document that they had done their work conscientiously. As the Archbishop of Toulouse and the Bishop of St Papoul had done before them, they had followed the indefatigable Riquet into the forest fastnesses of the Montagne Noire like so many pilgrims seeking grace and, like their forerunners, they were convinced. They estimated that Riquet's proposed feeders would draw upon a catchment area of eight square leagues (approximately 85 square miles) having an average annual rainfall of 26 inches. While they were satisfied that this would prove adequate for most of the year, they doubted whether the supply could be maintained in time of summer drought. But they whittled Riquet's proposed fifteen or sixteen storage reservoirs down to two by recommending the damming of the valleys of the Lampy and the Rieutort.

The Commissioners had also visited a spring at Naurouze, known as the Fontaine de la Grave, which Riquet had pinpointed as being the exact spot to which he intended to bring his proposed main feeder channel, the *rigole de la plaine*. It was situated near a huge pile of boulders which marks the summit of the Col de Naurouze. Like all such geological phenomena, these Stones of Naurouze as they are called, became the subject of legend, being attributed to the work of giants or the Devil himself. In this case it was a giant who, conveying stone from the east for the building of Toulouse, tired on reaching the summit of the pass and dropped his heavy burden. Thanks to Pierre Paul Riquet, an equally picturesque myth has grown up around the Fontaine de la Grave. The Commissioners gravely noted in their report that 'when it rains, half the water goes away towards Toulouse and half towards Narbonne', concluding with a blinding glimpse of the obvious – that if water was

brought to the spot it would do likewise. Legend has it that
Riquet, on earlier observing the same simple fact, had
exclaimed 'Eureka', and at once hurried off in the direction
of the Montagne Noire. The whole problem of the Canal du
Midi had been solved in a single flash of inspiration. This is
the equivalent of our legend that James Watt 'invented' the
steam engine by observing the lid rising on a boiling kettle.
As Riquet and Watt would have been the first to point out,
invention is never as simple as that.

On the subject of the canal itself the Commissioners
reached the same conclusion as Riquet and Boutheroue –
that rivers were best avoided. They recommended that,
ideally, a still-water canal should be built all the way from
Toulouse to the Mediterranean. On the subject of its
Mediterranean terminus, however, the Commissioners had
ideas of their own. They had inspected the ports of La
Nouvelle at the mouth of the Robine, and that of Agde on
the river Hérault, and had decided that they were too old
and that neither would provide an adequate terminal for
such a grand work. Instead, they proposed building a completely new Mediterranean port at the north-eastern
extremity of the narrow isthmus dividing the Étang de
Thau from the sea. With this project, Riquet would have
nothing to do. 'I have no knowledge of the question,' he
told Colbert, 'And come back to the subject of the canal.'
But he was not dismayed by the proposal, pointing out that
it would be easy to extend his canal from the Aude/Robine
junction near Narbonne, across the Étang de Vendres (a
drained salt lake) and over the rivers Orb and Hérault to
join the south-western end of the Étang de Thau.

This last recommendation of the Commissioners was
accepted with alacrity and the first stone of a new harbour

which was destined to become the Port of Cette (now known as Sète) was laid on 9 July 1666, six months before work on the canal was begun. One reason for this delay was that the Commissioners counselled caution on the subject of the feeder channel to the summit level, on which the success of the whole venture depended. They advised that as a preliminary step a small ditch, two feet wide, should be dug along the line of the proposed main feeder to the summit so as to prove beyond question that the levels were right and that the waters of the Sor would really flow to Naurouze as Riquet had claimed. So confident was Riquet that he sportingly agreed to dig such a ditch at his own expense and claim reimbursement only if he succeeded. He was authorised to carry out this work by letters patent dated 27 May 1665.

In view of the difficulties with heavy earthworks which he clearly anticipated when he had written to Colbert in October 1664, this was a brave gesture on Riquet's part, but it paid off. It is clear from the jubilant letter he wrote to Colbert on 31 July 1665 that he had found a longer but cheaper route for the feeder by carrying it along the contours, thus avoiding costly earthworks. He wrote:

Many people will be surprised at the very short time it has taken me and the little money I have spent. As to success, it is certain, but in a new way that no one had ever thought of. I may be counted among those, as I may assure you that the way I am proceeding now had always been unknown to me, whatever my efforts to discover it. The idea came to me while at St Germain. I thought of it a lot and, although far away, my dream turned out to be right, once on the spot. The level confirmed what my imagination had suggested two hundred leagues away. Through such a novelty, I rid my work from every fill, bank and cutting and I take it on the ground surface, across hollows and dales by natural gradients, so that I make the work easy and the thing easy to maintain, and I save about 400,000 livres of expense on the great deviation feeder, the estimated cost for those fillings, banks and cuttings, and the long time it would have taken to collect the materials and for the construction.

The result of this rethinking, as we can see today, is a classic early example of contour canal cutting.

The tenor of Riquet's letter suggests that by the time he wrote it his small trial ditch was well on the way to completion, yet all the authorities agree in stating that this crucial exercise which meant so much to Riquet was not completed until October. The explanation is that Riquet's enthusiasm and confidence were such that he began to enlarge the ditch into a full-scale feeder channel before it had reached Naurouze. This is made clear from the following letter which he despatched to Colbert on 28 September:

It is not without reason there is a saying: 'When you start eating, you soon get hungry.' Its truth is found in the execution of my venture. I had started with a small experimental feeder, and I carry on with one that could be a main deviation channel in a lesser undertaking than mine, as in truth, the amount of water I am taking to Naurouze would nearly suffice for the upkeep of a canal such as the one of Briare. The trouble is that with the increase in work and the rain falling here for the last fortnight, the completion is deferred and the expense increased, so that I shall not have finished before the end of October next approximately, and that the cost will come to the neighbourhood of 50,000 livres. But, my Lord, the least favourable and most incredulous will be compelled to admit through such a material demonstration that what I have done is a fine thing. Few believed in success, and now that it cannot be doubted, most say that what I have done amounts to a miracle, that it couldn't be done without God's help or the Devil's cooperation. I am the first one to agree and, on the whole, I shall be given my due when people say of me that I have some natural gifts but am no artist, and no magician.

Before the end of October 1665, as Riquet had promised, his work was completed and the waters of the Sor, diverted from their course, flowed obediently into the Fontaine de la Grave at Naurouze to convince the most sceptical that a canal between the two seas had become a practical proposition. Then as now, the machinery of government, on the pretext of proper caution, always prevaricates when faced with the need to make an important decision. It snatches at any pretext to appoint another Commission or Committee of investigation. It must have seemed to Riquet that, when he had successfully completed his feeder channel, the case for the canal had been proved beyond further doubt. But there were still vital questions to be debated. First, how the project was to be financed, and secondly whether the canal should be owned and administered by the state or privately. These matters were considered by the King's Council and it was finally decided that construction would be financed on a tripartite basis by the central exchequer, by the Province of Languedoc and by the undertaker himself. The acquisition of the necessary land would be the responsibility of the state, while the proceeds of the salt tax in Languedoc would be used to defray the cost of construction. In return for his contribution, the undertaker and his successors would be granted the canal as a fief and enjoy the profits from tolls. For, the Council decided:

. . . a piece of work which requires a constant attention and incurs daily expenses could not be without inconvenience left to a public administration, and that it was more profitable and safer to leave the management to an individual, to give him its ownership, interest him in the maintenance of the thing, and put the public interest under the safeguard of the private. Such an arrangement ensured the strength, the upkeep and the improvement of the Canal. No interruption due either to financial difficulties or State misfortune was to be feared.

So matters dragged on for another year until, in October 1666, King Louis XIV issued an Edict announcing the construction of a canal between the two seas.[1] Its builders and their successors were granted the ownership of the canal in perpetuity with exemption from taxes on the property and the exclusive right to build houses, mills, docks and warehouses on the canal banks, and boats for carrying merchandise. They were also to enjoy the hunting and fishing rights and the right to collect tolls, but not to fix the scales of charges. They were also to appoint judiciaries and twelve guards, who would be entitled to wear the King's livery, to collect the tolls and to enforce the by-laws.

Following this Edict, letters patent were granted to Riquet on 18 November and registered at Montpellier on 7 March following. Although by this time the work of building the new port of Sète was actively proceeding, the question of the route for the eastern section of the canal had still not been settled. Hence the patent authorised Riquet to construct the western section only from the Garonne at Toulouse to the river Aude at Trèbes, a riverside village a little to the east of Carcassonne. The patent charged Riquet to complete this and to render all to perfection in eight years, commencing January 1667, for the sum of 3,360,000

1. The text of this document is translated in full as Appendix A.

livres. So the canal conceived by Leonardo da Vinci 150 years before was about to be translated into fact at last. In his sixty-second year, Pierre Paul Riquet set about a mammoth undertaking which would have daunted most men of half his age.

[THREE]

Construction of the Feeder System

Having been given the go-ahead at last, Riquet wasted no time. By January 1667 two thousand labourers were already at work and this was soon augmented, for on 15 March Riquet reported to Colbert that the number had already been doubled. At the peak of construction a labour force of more than 12,000 was under Riquet's command including 600 women who were recruited owing to a shortage of man-power. While the men excavated the spoil, the women carried it away in baskets on their heads. Riquet established an admirable system for the control of this vast labour force, dividing it into as many as twelve divisions, each under the supervision of an engineer or 'inspector general', while the workmen under him were organised into gangs of fifty, each of which was responsible to a foreman. In addition, a team of seven surveyors was kept permanently occupied.

Work began first on the water supply system in the Montagne Noire; work on the Toulouse–Naurouze section of the main canal began a little later. For the sake of clarity the history of this all-important feeder system will be dealt with first.

Général Andreossy states in his book that it was his great-grandfather who first proposed damming the Lampy valley in the Montagne Noire, and he reproduces in his book a plan of the proposed Lampy reservoir which, although it

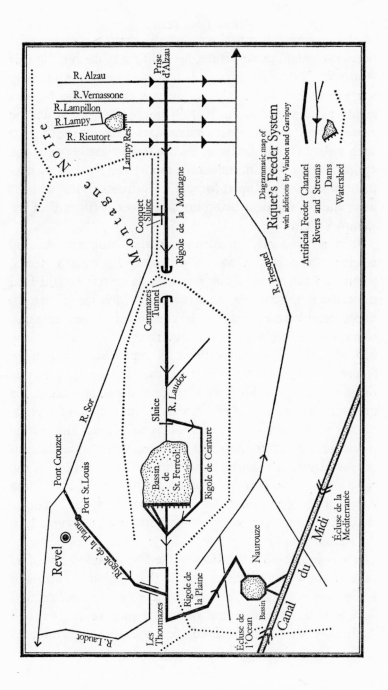

Diagrammatic map of
Riquet's Feeder System
with additions by Vauban and Garripuy

Artificial Feeder Channel
Rivers and Streams
Dams
Watershed

bears no legend or signature, he claims was the outcome of Andreossy's original survey. It may well be that Andreossy suggested this and made his survey at the time he was acting as surveyor to the Royal Commission, for it will be remembered that the Commissioners recommended that such a dam should be built. But at some time between the appearance of the Commission's report and the commencement of the work, Riquet (or was it Andreossy?) had a better idea, and for the time being nothing more was heard of the Lampy scheme.

It was evidently decided that the proposed Lampy reservoir was not adequate to meet the canal's needs, particularly in view of the fact that its waters would feed the canal indirectly via the river Sor, which meant that an unpredictable amount would be consumed as compensation water to the millowners on that river. What was clearly needed was a reservoir capable of replenishing the main feeder, or *rigole de la plaine* as it was called, directly. This was the thinking that led Riquet to propose the construction of a great earth dam across the mouth of the valley of Laudot stream at St Ferreol in the Montagne Noire. This valley extends from east to west and the stream that issues from it pursues a westerly course until it intersects the route of the *rigole de la plaine*. It then turns north to join the Sor below Revel. This meant that the Laudot would form a natural channel to convey water from the proposed reservoir into the feeder.

It was found that at St Ferreol the Laudot flows out of its valley over a natural sill of granite that extended unbroken across the valley floor, and it was this providential natural feature that fixed the line of the dam. As Riquet and Andreossy evidently realised, for the security and per-

8 Les Thoumazes: site of old lock and junction of feeder from St Ferreol (right) with the Rigole de la Plaine

9 Naurouze: end of Rigole de la Plaine and the Watermill

10   The Riquet Obelisk on the Stones of Naurouze

11   The old lock house, Naurouze

manence of a gravity dam (i.e. one that depends for its stability entirely on its weight) it is essential that its footing be absolutely impermeable and, ideally, anchored in bedrock throughout. Most failures of gravity dams that have occurred either before or after this date have been caused by the impounded water percolating through defective strata beneath the dam and so undermining it.

The foundation stone of the St Ferreol dam was laid on 15 April 1667 and it was completed in four years. The first dam ever to be built to supply a navigable canal, it was by far the greatest single work of civil engineering undertaken during the building of the Canal du Midi. It can be referred to in the present tense because, in its 300 years of life, it has required no major repairs but stands to this day, still performing the function for which it was built. It is an impressive monument to the genius and daring of Pierre Paul Riquet and his engineers. To anyone who walks along the broad crest of the St Ferreol dam today, gazing out over the great expanse of tree-girt water that it created, the fact that it was built so long ago seems scarcely credible.

Beyond the Pyrenees, whose snow-capped summits, visible in clear weather from the slopes of the Montagne Noire, wall the southern horizon, some very remarkable masonry dams had already been built in Spain by this date, the most notable being that at Alicante which was completed in 1594 and reached a height of 134 ft 6 in, a figure not surpassed for nearly three centuries. But, magnificent achievement though it was, the Alicante Dam fills a narrow gorge so that its crest length is only 262 ft, a figure which dwindles to a mere 30 ft at its base. For the sheer volume of material used in construction it cannot be compared with St Ferreol, which has a crest length of 2560 ft, rises to a maximum height of 105 ft above the bed of the Laudot and has a base thickness of more than 450 ft. The total volume of material used is 208,000 cubic yards, all of which was placed by hand, the earth fill being carried to the site in baskets, much of it by women, who were paid at the rate of a penny per load. It consists of a mass of well-compacted earth and stone sandwiched between two retaining walls and one central core wall, all of massive masonry securely founded on the granite sill of the valley. The downstream wall is 60 ft high and tapers in thickness from 30 ft at the base to 17 ft at the top. From this the earth fill, now thickly planted with trees, slopes upward towards the core wall which extends to the full height of the dam, is 17 ft thick and built with counterforts on its downstream side. An unusual feature of the dam is that the 48 ft thick upstream wall and the sloping earth fill between it and the core wall is completely submerged when the reservoir is full, so that what the visitor sees is the top of the core wall forming a low parapet behind which the earth fill is levelled off to make a broad promenade along the crest of the dam. To prevent water percolation, the slope of earth fill on the upstream side is covered with a layer of clay more than six feet thick. In addition, the lower part of the slope is faced with stone slabs one foot thick, to check wave erosion.

Far below, following the original bed of the Laudot and penetrating the base of the dam, is a stone vaulted tunnel which was called dramatically the '*Voûte d'entrée d'Enfer*'. The mouth of this tunnel can be seen at the base of the downstream wall. At its upstream end it communicates with the bottom of a vertical masonry shaft or well that descends through the waters of the reservoir and has a series of holes pierced through it at vertical intervals so that, no matter how its level varies, water may be drawn off through the tunnel. In addition there is a large sluice way in the base of this outlet well which, when opened, enables accumulated silt in the reservoir to be scoured away by water pressure through the tunnel. Many early reservoirs have completely filled up owing to the lack of any such adequate provision in the dam for flushing the silt away. This arrangement of vertical well, sluice door and low level outlet tunnel is an almost exact model of that evolved in the earlier Spanish dams, so it seems highly probable that either Riquet or Andreossy had visited Spain and studied the state of the art in that country before designing St Ferreol. The outlet of water through the '*Voûte d'entrée d'Enfer*' is regulated by three gigantic brass cocks situated in a subterranean valve house 90 ft below the dam's crest. Access to this is obtained through a descending passage way with a gated entrance higher up in the face of the downstream wall. The silt sluice is situated 10 ft below this valve house.

At the north end of the dam there is a generous overflow spillway because, as its designers realised, in the case of an earth dam of this type it was obviously important that water should never be allowed to overtop the dam crest. Beside this spillway are two high-level sluices which can be used instead of the low-level outlet cocks so long as the reservoir remains moderately full. When filled to capacity, the St Ferreol reservoir contains 180 million gallons of water (7 m cubic metres).

In its day the St Ferreol dam was the greatest civil engineering work of its kind in Europe. It presaged that mastery over his environment which man was destined to win in the course of the next two centuries, a conquest

marked by the growth of a brutal arrogance and insensitivity
such as we do not find at St Ferreol. For here we can see
that the Renaissance engineer did not allow himself to be
governed solely by the narrow motive of commercial gain
but, for the greater glory of his country, ensured that his
works would enhance its natural beauties. Although in the
seventeenth century the valley of the Laudot was in deep
country, far more remote and inaccessible than it is today,
like a skilled and sensitive landscape architect Riquet planted
trees in the gorge beneath his mighty dam, laid out walks
beside the rushing streams and led the overflow from the
spill weir down into these deep green shades by a spectacular series of waterfalls and cataracts. And, as a final touch,
he took advantage of the head of water available to place a
fountain near the foot of this man-made cascade. This shoots
a single powerful jet of water sixty feet into the air whence it
falls as an iridescent white curtain, swaying like a ghostly
tree amid the green. This fountain symbolises that instinct
to combine utility with beauty and grandeur, engineering
with art, that informs the whole design and construction of
the Canal du Midi. In this country, certain of the works of
Telford, Rennie and Brunel convey the same conscious sense
of splendour, of great occasion, but after Brunel the inspiration was lost and no engineer could speak the grand language of the Renaissance.

While this great dam was building at St Ferreol, other
gangs were at work on the southern slopes of the mountain,
digging Riquet's *rigole de la montagne* from the Alzau stream
near Ramondans wood, collecting the waters of other
tributary streams along the way and finally, at Conquet,
discharging through a cutting and down a 25 ft cascade
into the headwaters of the river Sor. So small was the con-

tribution of the Laudot stream to the St Ferreol reservoir that it became empty by the latter part of the summer and, once empty, it took anything from two to three months to fill it again. Consequently it was very soon decided to augment the natural flow of the Laudot by extending the *rigole de la montagne* for a further 4½ miles. This work was carried out in 1686–87 under the direction of the great military engineer Marshal Sébastien Vauban (1633–1707).[1] Vauban carried the extended feeder through the ridge on which the village of les Cammazes stands by a 132 yard tunnel 9 feet in diameter followed by a long cutting. This Cammazes tunnel has impressive portals of monumental stonework, the eastern one carrying a tablet inscribed with distances. There is a railed walkway through it for inspection and maintenance purposes. Where the feeder emerges from its cutting beyond the tunnel it is almost directly above the source of the Laudot and it descends the valley to join this stream by a series of cascades, carefully graded and stepped so as not to cause scouring. The water then follows the natural bed of the stream as far as the reservoir. At the head of the reservoir, however, the Laudot flows into a small chamber having two discharge sluices set at right-angles to each other, the one discharging into the reservoir and the other into a second artificial channel known as the *rigole de ceinture* (the belt feeder) which skirts the southern shore of the reservoir to fall into the Laudot stream below the dam. The purpose of this is to enable the feed to the reservoir to be diverted if the latter has to be emptied for scouring purposes. Also, of course, it can be used to relieve

1. It appears that Riquet originally proposed extending the mountain feeder to join his *rigole de la plaine* at les Thoumazes, but that when it was decided to build the St Ferreol reservoir this plan was abandoned.

the spillway at the dam should abnormal rainfall cause heavy flooding.

From below the St Ferreol dam, the feeder water, together with that discharged from the dam, follows the natural course of the Laudot for three miles to High Mill (Moulin Haut) where it is again diverted into an artificial channel which, in another mile, makes a junction with the *rigole de la plaine* at Les Thoumazes, just beside the road from Castelnaudary to Revel (Route Nationale No. 264). Immediately below the junction of the two feeders is a stone chamber equipped with flood paddles that allow excess water to be discharged into the Laudot, which here turns northwards to join the Sor whereas the main feeder channel heads southwest on its circuitous route along the contours towards Naurouze. The following heights above the summit pound of the canal at Naurouze will convey some idea of the gradients involved on this feeder system: they are:

Alzau Intake: 1500 ft

Base of St Ferreol Dam: 312 ft

Junction of feeders at Les Thoumazes: 88 ft 6 in

When the extension of the *rigole de la montagne* was made by Vauban, a discharge sluice was provided at Conquet so that a proportion of the water can still be released into the Sor through the original channel provided by Riquet.

This feeder system and the St Ferreol reservoir between them constituted the sole water supply to the canal between Toulouse and the Fresquel intake near Carcassone for nearly a hundred years. It was estimated in 1818 that the feeders alone yielded an average flow per twenty-four hours of 4,251,500 gallons (19,325 cubic metres), a figure that the St Ferreol reservoir boosted to a total of 79,403,500 gallons

(360,925 cubic metres).[1] Nevertheless, it seems that Riquet's original plan for a multiplicity of storage reservoirs in the valleys of the Montagne Noire was not so wide of the mark after all. For according to de Pommeuse, despite the size of the St Ferreol reservoir, during the greater part of the eighteenth century the canal had to be closed during the period of summer drought for an average period of eight weeks each year owing to water shortage. Advantage was taken of these annual closure periods to carry out any necessary maintenance works.

The fact that the line eventually adopted for the canal by-passed the city of Narbonne was a source of bitter complaint, as we shall see. That city urged the construction of a short junction canal, falling from the main line of the Canal du Midi to link up with the Aude and the Canal de Robine. The fact that such a link was not provided until 1776 was entirely due to fears that it would aggravate the water shortage in summer. For although in this area supplies from the summit were supplemented by intakes from the rivers Fresquel, l'Orbiel and Cesse, it was feared that the proposed branch would draw too much water away from the main line. It was the eventual construction of this Narbonne branch, coupled with an increase in traffic, that led to a decision to dam the Lampy valley as the Royal Commission had originally recommended over a hundred years before. De Pommeuse states that the additional reserve created by the Lampy dam not only supplied the needs of the Narbonne branch but also reduced the length of the annual summer stoppage from eight weeks to two. This is a somewhat surprising statement in view of the fact

1. Figures quoted by the Chief Engineer of the Canal, M. Clauzade, to Huerne de Pommeuse (see Bibliography).

that the capacity of the Lampy reservoir is only a third of the St Ferreol.

The Lampy dam was built between 1777 and 1781 and it is still in use. It was designed by the then chief engineer of the canal M. Garripuy, and was built under his direction. Unlike that at St Ferreol, it is a masonry buttress dam, the second of this type to be built in Europe. The first, the Almendralejo dam in northern Spain, thirty-two miles north of Badajoz, was completed thirty years before. This suggests that Garripuy, like Riquet before him, may have been aware of Spanish practice. But if he did base his design on Almendralejo, then he was not entirely successful for although both still stand, the Lampy dam is the less satis-factory of the two.

The dam consists of a wall of granite blocks, hewn locally. It tapers in section with one step on the upstream side and is 17 ft thick at the crest, broadening to 37 ft at the base. It is 53 ft high, its crest length is 385 ft and it stands upon a masonry plinth, 46 ft thick and 6 ft 6 in high, which is anchored in the bedrock. This plinth projects beyond the downstream face of the dam wall and so also acts as a base for the ten masonry buttresses which support the dam. These buttresses taper from base to crest, but not generously, and even to the eye of a layman they appear to give scant support to the dam wall. There are three outlets at different levels, each controlled by a sluice gate and there is also a spillway.

According to Général Andreossy, when the Lampy reservoir was first filled, water began to leak through joints in the masonry of the dam. To counteract this, quantities of slaked lime were thrown into the water in the hope that the lime would find its way into the joints and seal them.

Apparently this expedient was successful, but not permanently so for today the dam wall leaks considerably. At some subsequent date, in an effort to strengthen the dam, dozens of holes were driven vertically through the wall from the crest into the foundations. Into each an iron bar was inserted, grouted into the base and then tensioned by tightening a large nut on the crest. On a recent visit, Dr Norman Smith found the sight of dozens of nuts protruding from the crest of the dam wall 'even more alarming' than the leakage through the face.[1] The Lampy reservoir has a maximum capacity of 506 million gallons (2·3 million cubic metres), or under a third of the capacity of St Ferreol.

Over the years various proposals were advanced to increase the water storage capacity in the Montagne Noire. A dam 82 ft high was planned in the valley of the Alzau at Arfonds which would create a reservoir of a capacity greater than that at St Ferreol. Another plan was to dam the valley of the Sor to impound a reservoir of 9 million cubic metres capacity, and yet another proposal was to double the size of the basin at Naurouze. None of these plans materialised, so there was no further increase in water reserves until 1956 when a new reservoir was completed near the village of Les Cammazes. Although primarily intended for potable water supply and irrigation, this is connected to the canal feeder system and the Canal Service has an option to draw from it up to 880 million gallons (4 million cubic metres) a year.

The greater part of the *rigole de la plaine* from the Sor intake at Pontcrouzet to Naurouze is almost as large as some of our narrow canals, being 20 ft wide at the surface while its average depth of 6 ft extends to a bottom width of 12 ft. This was not always so. Soon after work on the feeder

1. *A History of Dams*, p. 163 (see Bibliography).

began in 1665 Riquet enlarged his original 2 ft trial channel to a width of 12 ft at the surface and a base width of 6 ft. In 1668, however, he decided to make use of the feeder in order to transport stone and other materials needed for the construction of the canal. For this purpose a loading basin, 126 ft square and 10–12 ft deep, was built near Revel and named Port St Louis. At the same time a large octagonal basin was provided at Naurouze. This was 9 ft deep and measured 1200 ft in length and 900 ft in width. Between these two terminals the feeder was widened to its present dimensions and a fleet of small boats was built measuring 24 ft long by 6 ft beam. Because the feeder's gradient was too great for traffic, fourteen locks were built upon it. Being originally intended merely as a temporary expedient these were simple affairs with sloping turf sides, timber piled at each end where the gates were hung. Entrance width was between 6 and 7 ft and the length between gates was 120 ft, sufficient to pass five boats at a time.

Originally Riquet, ever anxious to secure additional water supplies, designed his feeder to take in various streams which, rising in the Lauragais and flowing south-eastward, intersected the line of route, a notable example being that which has its source near the village of St Felix. It was soon found, however, that in time of flood such streams not only caused the feeder to overtop its banks but brought quantities of silt into its bed. So all these streams had to be conveyed in culverts under the feeder or, in one case, over it in a small aqueduct. This was a lesson which Riquet and his engineers unfortunately failed to learn for, as we shall see later, the same trouble was experienced on a far greater scale on the main canal and had to be rectified by Riquet's immediate successors at considerable expense.

Having made his feeder navigable for his own purposes, Riquet next proposed that this should become permanent and even suggested a possible extension to Castres, presumably to placate its citizens for the failure of their original scheme.[1] But although this extension never materialised, he envisaged his Port St Louis at Revel becoming a centre for the shipment of grain from the area, cargoes being transhipped to larger craft at Naurouze. In association with this plan a most grandiose scheme was devised for Naurouze. A new town was to be laid out around the basin with houses modelled on those of the Place Royale in Paris and complete with churches, a monastery and dockyards. Covered promenades would extend to the quays beside the basin which would become the centrepiece of this early example of town planning. In the middle of the basin there would be set a gigantic fountain depicting Louis XIV, his foot upon a globe, riding in a chariot drawn by sea-horses.

Regular trade on the feeder never proved a success and soon lapsed. In 1705 a man named Laval tried to re-establish navigation by making it easier for loaded boats to pass each other, but his efforts failed and navigation was abandoned in 1725. The proprietors of the Canal du Midi then acquired the navigation rights and planned to widen the feeder and to build new and larger locks, but they gave up their plan, fearing that such an enlargement might jeopardise the waterway's primary function of supplying their main canal

1. In his letter to Colbert of October 1664, he had suggested that his main canal 'could attract smaller canals coming from various towns of the province. For none is there any insurmountable natural obstacle. The Albigeois could be linked through the river Tarn, the Castres country through the Agoût, etc.'

with water. Finally, towards the end of the eighteenth century, a Toulouse engineer named Meyer proposed a new canal from Naurouze to the Garonne via the rivers Agoût and Tarn. This was no more than a revival of the ancient proposal of 1662, but whether he was aware of this or not, nothing came of Meyer's project.

Traces of three masonry lock chambers are now the only surviving evidence that Riquet's feeder was ever used for trade. The first of these is situated immediately below the junction of the *rigole de la montagne* at les Thoumazes. The obvious lock cottage beside it is now occupied by the man responsible for operating the flood paddles there which discharge into the Laudot. There is a modern flow-meter beside the old chamber and also a flood marker dated 1 June 1871 as a reminder that, although floods may be rare, when they do come they can be prodigious. Incidentally, large-scale maps (1/50,000) still mark *Écluse* at this point.

The second masonry lock chamber can be seen at Naurouze where the feeder formerly entered the basin. The waters of the feeder now drive a water turbine below the mill which stands beside the old lock chamber. The present mill building bears an 1844 date label, but it doubtless occupies the site of a much older mill that took its power from the lock by-pass weir. The old chamber now has fixed sluices which regulate the flow that by-passes the mill. At the opposite, western, end of the basin site, traces of a third lock chamber, again with an obvious lock house beside it, can also be seen. It was used to enable small boats from the feeder canal to enter the Canal du Midi in a westerly direction. Whether these masonry locks were originally built by Riquet at the time he decided to make navigation on the feeder permanent, by Laval when he attempted to restore it, or by the proprietors of the canal before they had second thoughts, is not known.

As for the great basin at Naurouze, its fine stone copings crumbled and it gradually decayed. Yet even in its decay it served a useful function for many years by acting as a vast settling tank for the silt brought down by the feeder, which would otherwise have been carried into the summit level of the canal. As a result it has completely filled up and grown over. But its dimensions are preserved by the feeder channel which now completely encircles the area like a moat, to the mystification of strangers who do not know the story of Naurouze.

The watermill and the house beside the old entrance lock are the only buildings on a site where once a thriving new town was planned. Yet tall trees, lush grass and the rippling waters of the feeder make this a cool green oasis in the sun-bleached and arid Col de Naurouze. Moreover, it is not, as was originally planned, Louis XIV whose memory is perpetuated in this place but that of Pierre Paul Riquet. In 1825 his descendants, the Riquets de Caraman, purchased the famous Stones of Naurouze and erected on the summit of that formidable pile of boulders a slender obelisk bearing a portrait medallion at its base to commemorate their famous ancestor.

In 1837 Général the Duc de Caraman caused a great inscribed block of granite to be set up to the memory of Riquet at the high starting point of his mountain feeder, the Prise d'Alzau which, like the valley of the Laudot below the St Ferreol dam, was landscaped with trees and walks. These two memorials are fittingly placed for, as Riquet correctly foresaw, the success of his great canal from sea to sea depended entirely upon the reliability and adequacy of this elaborate feeder system consisting of forty miles of feeder channels and the mighty dam at St Ferreol.

Despite de Pommeuse's pessimistic statement, there appears to be no record of the canal being closed because of shortage of water since the completion of the Lampy reservoir and there has certainly been no such closure within living memory, although the canal came very near to it during the exceptionally dry summer of 1948. In that year all boatmen were ordered to reduce their laden draft and to 'wait turns'[1] at the locks, while lock-keepers were instructed to stop all leaks through the lock gates with cinders. But with the extra reserve provided by the new reservoir at Les Cammazes it is estimated that there are sufficient supplies of water to feed the canal for four months without a drop of rain falling. To sum up, Riquet's feeder system was an object lesson from which subsequent British canal engineers singularly failed to profit. With its later additions it is still an object lesson today to those who glibly argue that our own canals should be enlarged to Canal du Midi dimensions or larger without considering the water supply problems involved.

1. 'Waiting turns' is a system of working whereby a barge proceeding 'uphill', on coming to a full lock, is not allowed to proceed but must wait until the lock is used by another craft proceeding in the opposite direction and vice versa.

[FOUR]

Building the Canal
1667-81

The route planned for the canal between Toulouse and
Naurouze lay along the southern slopes of the valley of the
little river Lers. Riquet and his engineers encountered
considerable opposition from the landowners concerned
who argued that the new waterway should be confined to
the floor of the valley where the land was of lower value.
This was merely a revival of the old idea of making the
river Lers navigable and Riquet would have none of it.
He argued that the land in question was useless to him
because it was subject to heavy flooding and was at all times
boggy, offering no firm foundation for a towing path.[1]
With the power of the King and Colbert on his side, in this
dispute as in so many others, he won the day.

It must be remembered that Riquet had now been placed
in a position of unique power in Languedoc. Not only had
he been charged with the responsibility for building the
canal but also, by virtue of his royal appointment as commissioner general of taxes, with the power to extract, from
friend and opponent alike, the money to pay for it. It was
no wonder that in carrying out his mammoth undertaking

1. The valley has since been drained by diverting the waters of the
Lers from their old bed into a straight drainage dyke.

Riquet encountered so much jealousy, envy and enmity. The fact that he committed the whole of his considerable private fortune to the undertaking cut no ice with those who, despite their opposition, were compelled by Riquet's team of tax gatherers to contribute to it.

This initial dispute may have had something to do with the fact that after the commencement of work on the St Ferreol dam, seven months elapsed before work on the canal proper was put in hand. It began with a ceremony at Toulouse where two foundation stones were laid in the wing walls of the entrance lock from the Garonne on 17 November 1667. Bronze medals and plaques bearing flowery Latin inscriptions were struck for the occasion. Exactly a year later, this first lock was completed and formally opened with even more pomp and circumstance amid scenes of great popular jollification including processions through the streets of the city. Many of the workmen engaged on the St Ferreol dam came down to Toulouse for the occasion, while the 'three musketeers' of the canal, Riquet, Andreossy and Campmas, stood side by side on the lock wall as the Archbishop of Toulouse, Mgr Anglure de Boulement, blessed the work. When he had done so it was Pierre Campmas who drew the first sluice which admitted water to the new chamber.

The first design of lock adopted had the normal rectangular chamber and is said to have been based on those built on the Tarn Navigation. The prototype at Toulouse was of considerable depth as it was important that the octagonal transhipment basin known as the Port de l'Embouchure above it should be beyond the reach of the highest floods on the river. At the opposite end of the port from the entrance lock the Canal du Midi debouched, its mouth spanned by a

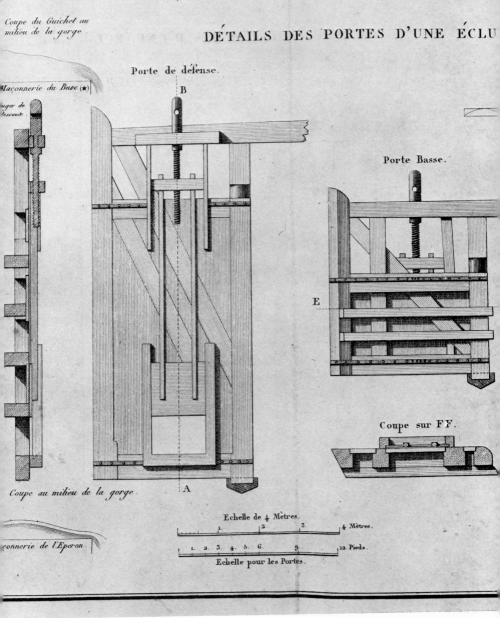

Coupe du Guichet au milieu de la gorge.

Maçonnerie du Busc (★)

...ger de ...essous.

Porte de défense.

B

Porte Basse.

E

Coupe au milieu de la gorge.

A

Coupe sur FF.

...connerie de l'Éperon.

Echelle de 4 Mètres.

1.   2.   3.   4. Mètres.

1. 2. 3. 4. 5. 6.   9.   12. Pieds.

Echelle pour les Portes.

Early type of timber lock-gate used on the Canal du Midi

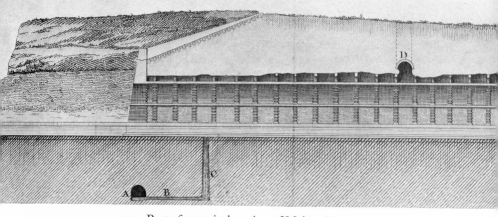

13　Part of a vertical section of Malpas Tunnel showing
one of the two vertical shafts at D, also the discharge
CB into the medieval drainage culvert A

14　The eastern (*above*) and western portals (*below*) of
Malpas Tunnel

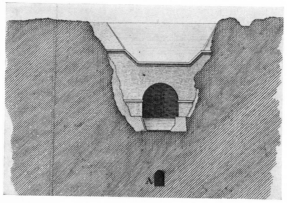

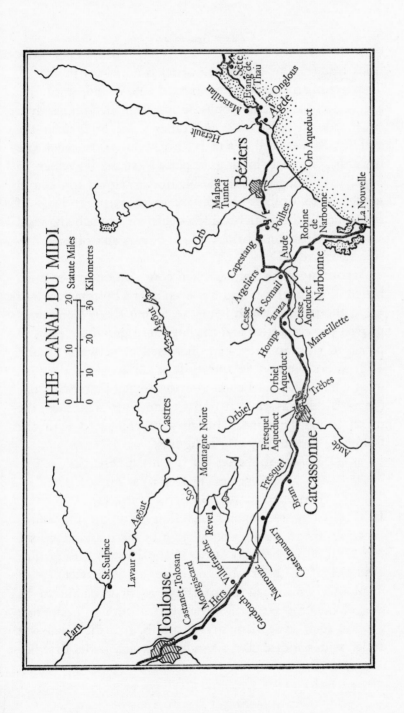

THE CANAL DU MIDI

brick bridge. The Toulouse district is noted for its bricks.
They are of an attractive texture and colour and, until this
age of ferro-concrete, Toulouse was still predominantly a
'rose red city'. For this and other canal bridges in the
vicinity, Riquet used this native brick, but as the canal rose
towards Naurouze, brick gave place to stone. For when he
began excavating the basin at Naurouze, Riquet announced
jubilantly to Colbert the discovery of excellent beds of
freestone. So the basin became a quarry, its stone being used
extensively for canal bridges, lock houses and other canal-
side structures.

According to La Lande, when construction of the canal
began its channel was dug 6 ft deep over a bottom width of
30 ft, the surface width being 56 ft, but these dimensions
proved too restrictive and the angle of slope too steep and
prone to slip. So they were increased to between 42 and
48 ft at bottom and 72 and 90 ft at surface depending on
local conditions. La Lande goes on to say that in his day
(1778) width was 32 ft at bottom and 60 ft at surface, the
depth being the same. This may be compared with the
following present day official figures:

> Width: 10 metres (33 ft 4 in) at base.
>
> 16 metres (53 ft 4 in) at surface.
>
> Permitted draft: 1·6 metres (5 ft 4 in).

It is obvious to anyone travelling through the canal,
however, that these are average figures and that, in fact, the
top and bottom breadth of the channel varies greatly. For
example, it is noticeably narrower in the vicinity of
Toulouse, which was presumably the first section to be
dug. Where the excavation was carried through permeable
ground such as sand or gravel, the bed of the canal was
sealed with puddled clay, a specific to which James Brindley

was later to become so addicted that Englishmen have come to believe that it was his invention.

On the eastern side of the summit particular precautions were taken to protect the banks of the canal and the lands bordering them from damage and inundation by flooding. A narrow step or berm was formed just below water-level and its upper surface planted with reeds to break the wash. As a further precaution, the banks were planted with trees. When the wash of a passing boat reveals them, it can be seen how effectively the meshed roots of these trees hold the banks and check erosion. Wherever the level of the canal was above that of the surrounding countryside, either upon one side or upon both, the material excavated from its bed was used to rear a continuous flood bank or *franc bord* (lit: free board) which enclosed both the canal and its towpath. As an additional safeguard, on the further side of this bank a deep ditch known as a *contre canal* was dug which carried away into the bed of the nearest stream any flood water which might overtop the bank. Wherever a stream was culverted beneath the canal, a massive masonry spill weir was constructed on the downstream side to discharge flood water into its bed. If the downstream side was also the towpath side, the latter was carried on arches over the sill of the weir. In addition, flood discharge sluices were frequently provided at such crossing points.

On a work so unprecedented, troubles and setbacks were inevitable, and one admires the way Riquet and his engineers seem to have kept their heads and tackled them resolutely. When some design fault became evident they did not tinker with it but ruthlessly scrapped it and started again. This may have been an expensive policy, but nothing less than the best would suffice for men working *pour la gloire* on a work which they intended should stand for all time. The first major disaster to occur was the collapse of the side-walls of one of the newly built deep locks. This caving inwards of lock chamber walls due to pressure from the surrounding ground, particularly if the lock was a deep one, is a trouble that many subsequent generations of canal engineers would experience. Which lock it was that failed does not appear to have been recorded – perhaps it was the original entrance lock from the Garonne. Nor do we know how many locks had been completed or were under construction at the time of the collapse. But presumably the canal had been completed from the Garonne to a point known as Dupérier because, at the beginning of 1670, Riquet warned Colbert that the completion of the canal between Dupérier and Naurouze would be delayed by his decision to modify the locks. In fact he had decided upon a completely new design.

The new lock chambers were to be oval on plan, being 36 ft wide at the midpoint and narrowing to 19 ft 8 in at the gates. Length between gates was 100 ft. Riquet told Colbert that he had settled upon this design of chamber in order to accommodate two *caponts* or *capons*, the type of vessel then trading on the Rhône,[1] which had a maximum beam of 16 ft. If this was so, it was either singularly fortuitous or the wily Riquet had decided to kill two birds with one stone. For the fact that his lock side-walls now took the form of horizontal arches meant that they were the better able to resist ground pressure. The sole disadvantage of the design was that, with straight-sided craft of the barge type, more water is consumed in lockage than would be the case with a

1. According to Jean Girou they were also used on the Robine at Narbonne.

lock having parallel sides, but Riquet envisaged that his canal would be used by vessels of seagoing form.

Riquet certainly ensured that such a disastrous lock collapse would never occur again. He greatly reduced the depth of the lock chambers so that their average fall was 8 ft and seldom exceeded 9 ft 6 in. Where the survey called for a greater fall than this he divided it between more than one chamber using intermediate gates, thus forming double locks ('risers') or multiple staircases as Cosnier had done on the Briare Canal. The side-walls were built with a slight batter and were 6 ft thick. The black volcanic stone from Agde, which had been used in the building of Agde's fortress-like cathedral, was preferred on account of its great durability, but where its use was not practicable owing to transport difficulties the locks were built with selected stone from quarries at Pesens or de Brezimes. This masonry was laid in mortar made by mixing lime with Pozzolana[1] brought from Civita Vecchia, near Rome.

Curiously enough, the example of Cosnier's ground sluices at the locks on the Briare canal was not followed owing, one can only assume, to the added cost and difficulty of building the necessary underground culverts. Nor, with six exceptions, were any spill weirs originally provided beside the locks, which meant that any excess water in the pounds poured over the upper lock gates. The exceptions were the locks or lock staircases at des Minimes, Matabiau, Bayard and Castanet on the western side of the summit and those at St Roch and Trèbes on the eastern side, where by-pass weirs were installed to drive mills.

1. Pozzolana is volcanic earth which, when added to lime, gives it hydraulic properties. It takes its name from Pozzuoli where it was first exploited.

The lock gates were of oak of the normal mitre pattern with balance beams. Each contained one sluice of large size drawn up by a vertical screw terminating at its upper end in a pierced boss through which a capstan bar could be inserted. The drawback to such an arrangement, as compared with ground paddles, was that unless the upper gate sluices were drawn judiciously the bows of ascending craft were deluged with water, a fact which meant that locking up was a slower proceeding than it might otherwise have been. Also according to de Pommeuse, owing to the erosive action of the water pouring through the sluice apertures, the average life of the timber gates was five years shorter than those on the Briare Canal.

Three things combined to upset Riquet during 1669. Like most men who undertake an unprecedented venture of this kind, he seems to have found the human difficulties much more trying than the physical ones. He became increasingly touchy, perhaps pardonably so, when anyone questioned his judgment in the conduct of the work or appeared to be undermining his authority. He was upset when a certain M. de la Feuille was sent down to inspect the work and to see whether it would be possible to economise in any way. Maybe this move was a result of Riquet's decision to replan the locks; in any case, with so much state money at stake it was not an unreasonable one. However, from an extraordinary letter he wrote to Colbert it appears that Riquet was convinced that M. de la Feuille was an enemy who had been sent to spy upon him. However, he eventually seems to have been convinced that this was not so and the two men parted on good terms.

Next, wearing his commissioner general of taxes' hat, Riquet found himself with what almost amounted to a civil war on his hands as certain parts of Languedoc revolted against the exactions of his tax collectors, on which his canal work depended. Apparently the men of Roussillon and the Miquelets were the most recalcitrant. To suppress this revolt Riquet acted ruthlessly and with evident success, explaining to Colbert that:

Murders are as common in Roussillon as bread and wine. A neighbour kills his neighbour, and a brother his own brother. No human power could prevent this kind of men killing each other, and you may therefore infer that the tax collectors are subject to the same fate. In this country, the tax collectors are always on the lookout; they kill just as they are killed; it is the only way to carry on their function.

Colbert was obviously appalled by this matter-of-fact statement and by other news he received of Riquet's actions in quelling the riots. To a protest against his severity, Riquet replied:

My Lord, I must say that the harm is not as great as you depict, and I have every reason to believe that in so doing the persons meant to injure me and to try at my own expense to make their pile by expelling from their jobs my employees who serve me well, to replace them by people who would serve me badly and conspire with the Miquelets.

These exchanges momentarily lift the lid off seventeenth-century Languedoc and make one marvel that a major civil engineering project could ever have been carried on at all against such a background of murder, violence and oppression. Doubtless Colbert marvelled too. The Occitans had ever been hotblooded and fiercely independent, mistrusting, and mistrusted by, the rest of France. Riquet understood them; Colbert did not.

The next thing was that François Andreossy, without Riquet's cognisance, published a three-volume map of the canal which he dedicated to the King. Now according to his great-grandson, Andreossy was at this time known as Director General of the canal. While this may or may not be true, we do know that he had overall responsibility for the detail survey work. This being so his action, though it may have been tactless, seems not unreasonable. But for Riquet it was betrayal; he had been stabbed in the back by his most trusted colleague. He at once wrote to Colbert:

I have been most surprised, my Lord, when I saw a certain map of the Canal, a contrivance of Mr Andreossy, my employee. This was done without my knowledge and I only heard of it later on, which was more cause for my displeasure, all the more so as the map is quite inaccurate and that he publishes thoughts which I meant to keep secret and did not want to carry into effect without having your advice and consent, as I already wrote to you. I mean the Naurouze basin column and many other things which have been disclosed against my will in this map. Which angers me as much as the manner in which it is presented, and it will make me circumspect in the future and more secretive towards said Mr Andreossy, and that possibly I shall use him no more.

It would be interesting to know what passed between Riquet and Andreossy at this juncture. All we know is that Riquet never carried out his threat of dismissal. Perhaps, when his hot temper had cooled, Riquet realised that Andreossy was indispensable to him, for we soon find the latter making detailed surveys to determine the route of the second section of the canal from Trèbes to the Étang de Thau. This must have been a salutary experience for him, for whereas his map had shown the low-level route that Riquet had originally proposed to the commissioners,

Riquet now asked him to survey a totally different line.

Despite such contretemps and Riquet's endless, worrying difficulties over money, the work went on. By the beginning of 1672 the canal was complete from Toulouse to Naurouze. The water from the main feeder was released westwards and in six days the canal was full as far as the Port de l'Embouchure, enabling three boats, the first carrying the Archbishop of Toulouse, to ascend to the summit. The first small boat had arrived along the feeder from Port St Louis in May 1668, and now through trade with Toulouse became possible. The first boat returned directly to Toulouse with the Archbishop, leaving the other two to load foodstuffs and barrels of wine. Thereafter boats plied regularly three times a week between Naurouze and Toulouse. Trading had begun on the Canal du Midi and the ebullient Riquet was naturally delighted. Moreover, work on the rest of this first section to Trèbes was progressing fast.

In 1673 M. de Bezons, King's commissioner in Languedoc, with the Bishop of St Papoul examined and reported upon the state of the works between Naurouze and Trèbes. They found the canal well on the way to completion. The locks between the summit and Castelnaudary had still to be completed, but only six miles of canal near Carcassonne had still to be dug. There was a particular reason for this omission. The river Fresquel flows into the Aude at St Pierre, about $2\frac{1}{2}$ miles downstream from Carcassonne. A ridge of high ground separates the valleys of the two converging rivers and since the canal follows the south side of the Fresquel valley, to have routed it into the city, which straddles the Aude valley, would have involved very heavy earthworks. The citizens of Carcassonne were anxious to have the canal, but when told that it would cost them 100,000 livres their enthusiasm cooled considerably. At the time the 1673 inspection was made, this issue was still being disputed, but shortly afterwards Carcassonne refused the extra money; so Riquet decided to follow the easier route, sticking to the Fresquel valley and eventually crossing that river on the level 50 yards upstream from the old Pont Rouge road bridge, half a mile from its confluence with the Aude. It was a decision which the citizens of Carcassonne would afterwards bitterly regret.

Riquet decided to adopt double and staircase locks solely for structural reasons. No water economy was achieved by such an arrangement. If, for example, a four-chamber staircase is used to overcome a difference of level of 36 ft, its water consumption is the same as it would be for a single very deep lock chamber. The only way consumption can be reduced is by the use of side-ponds. These consist of small reservoirs beside the lock chambers at an intermediate level into which half the water in a full chamber can be discharged through a ground sluice and there stored for use when the chamber is refilled. The inventor of this device appears to have been a certain Maître Dubie who, between 1643 and 1646, equipped a lock at Boesinghe, below Ypres on the Yser Lateral Canal. This had a fall of 20 ft, with two side-ponds. But if Cosnier or Riquet had heard of such an arrangement they did not adopt it. Consequently, wherever they had to build a considerable lock staircase they had to ensure that its use would not draw too much water away from the pound above.

The largest single work on the canal between Toulouse

and Trèbes is the quadruple staircase of St Roch at Castelnaudary. As this draws on a pound only 2½ miles long which
terminates at the single lock at Laplanque, Riquet realised
that additional water capacity would have to be created in
this pound to supply the St Roch staircase. Hence the
provision of the magnificent *Grand Bassin* at Castelnaudary,
an area of water spacious enough to contain several Atlantic
liners and a permanent asset to the town in a region comparatively waterless.

On the approach to Trèbes, another tributary of the
Aude, the river Orbiel, intersected the line of route and,
like the Fresquel, it was crossed on the level. This was done
by raising the river level up to that of the canal by constructing a masonry weir on the downstream side of the
canal. Because in both cases the towpath is on this side, it
was carried across the crests of the weirs on masonry
bridges, that over the Fresquel having eighteen spans.
Under conditions of normal flow such an arrangement was
perfectly satisfactory, especially as it enabled the canal to tap
additional water supplies. But, unfortunately, both rivers
rise in the Montagne Noir and are subject to sudden floods
of great violence. Although the weirs were of maximum
length and were equipped with additional flood sluices
operated from the towpath bridges, in practice they proved
incapable of coping with such a volume of flood water and
navigation had to be temporarily suspended. Moreover,
when a flood subsided it was usually found that it had
washed such a quantity of silt into the bed of the canal that
it had to be dredged out before traffic could be resumed. In
some cases single pairs of flood gates (*demi-écluses*) were
provided at such river crossings, which could be closed to
protect the canal in the event of exceptional floods, but it is

not known whether they were ever provided at these two places.

At Trèbes a triple staircase lock was built to bring the canal down to the level of the Aude, which here flows close beside it. Just over a quarter of a mile (500 metres) beyond the tail of this lock, a connection with the river was in fact made although it was provided for water intake purposes only, any idea of using the river below this point for navigation having been abandoned long before construction reached thus far. Sluices were provided here to isolate the river from the canal in time of flood and the masonry in which these were built may still be seen. But the only traffic to pass consisted of logs which at one time were floated down the river and through the sluices to be loaded into barges on the canal. The exact date when Riquet fulfilled his first obligation by completing his canal thus far is not known. In any case the question is purely academic because, unlike the Toulouse–Naurouze section, so far as commercial traffic was concerned it depended entirely on the completion of the second half of the canal from Trèbes to the Étang de Thau, which was already under way by the time the first half was completed.

Proprietorial rights over the Toulouse–Trèbes section, including the right to collect tolls, were formally granted to Riquet and his successors on 14 May 1668.[1] On 20 August in the same year he was authorised to construct the second section of the canal from Trèbes to the Étang de Thau, including work on the new port of Sète, in eight years for the sum of 5,832,000 livres. As before, the survey and estimates were checked by the Chevalier de Clerville.

1. Similar rights over the second section were granted on 16 January 1677.

Whether these were for the original route via a junction with the Aude near Narbonne, is not clear. It was certainly the route shown on Andreossy's map. But if this was so, Riquet soon changed his mind.

Ever since the idea of a canal between the two seas had first been mooted, the water supply question had always appeared to be by far the greatest obstacle. So it must have seemed to the layman, to Colbert and to the various inspectors and representatives of Church and state who buzzed about Riquet, that when traffic began to move between Toulouse and the summit, amply supplied with water from his feeder system, the battle had been three parts won. They were sadly disillusioned by Riquet's seemingly endless demands for money to complete the eastern section of the canal. Their feelings were exacerbated unwittingly by Riquet himself. He was growing old and suffered from bouts of ill health as a result of endless work and worry, and this did not help matters. With every month that passed the more obsessed did he become with the canal, the more obstinate and intolerant his attitude towards anyone who sought to criticise his policies or curb his expenditure. It was *his* canal which he alone knew how to complete. And, come hell or high water, it would be finished by him as he had planned it, not for the reward but for his own glorification. The incident of Andreossy and his map shows the state of Riquet's mind and the following extracts from letters he wrote to Colbert in 1670 and 1671 respectively are even more revealing:

As I am the inventor of this canal now being constructed in Languedoc and the one who studied it most, I am beginning to realise that I am also the one who knows it better than anybody else. My venture is the dearest of my children; I see the glory,

your satisfaction, and not the profit. I want to leave honours to
my children and I do not pretend to leave them any big amount
of money.

I am considering my work as the most cherished of my children,
which is so true that even with two daughters still to be settled, I
prefer to keep them at home for some time and use for my works
what I had destined to be their dowry.

What Riquet's family thought of his obsession, particularly
his two daughters, is not recorded. Fortunately, both eventually made good matches.

There is a certain naive and engaging honesty about
Riquet's letters so that despite his evident faults an endearing
and sympathetic personality shines through them. To Colbert, with all the cares of a whole country on his shoulders,
such letters must have seemed the tiresomely verbose and triv-
ial outpouring of an overweaning provincial obsessed with
his own importance. To Colbert, at St Germain, Languedoc
was a faraway and outlandish province and Riquet a
typically outlandish native who, through his patronage, had
acquired ideas above his station. But Colbert did not choose
to fall out with Riquet any more than Riquet chose to
dismiss Andreossy, and for the same reason. He was too
valuable. After all, the little upstart Occitan did appear to
be within sight of completing the greatest engineering work
in the history of Europe.

It is necessary to appreciate all this in order to understand
the bitterness of the opposition that Riquet encountered
over his choice of route for the eastern section of the canal.
It is also necessary to know something of the local topography. It will be remembered that Narbonne was the
original objective and that when the Royal Commission had

expressed themselves in favour of a new port at Sète, Riquet proposed extending a low-level canal to it from a point on the Aude near its junction with the Robine, a plan which would have kept Narbonne well in the picture. But Riquet had now had second thoughts and advocated an alternative route along the contours of the high ground, keeping the canal well above the flood plain of the Aude and the marshy *étangs* that lay between this high ground and the Mediterranean coastline. Finally, the canal would pass by a tunnel from the southern to the northern slopes of the Enserune ridge opposite the Étang de Montardy and so arrive on the west bank of the river Orb immediately above Riquet's native city of Béziers. From the east bank of the Orb to Agde and the Étang de Thau it would then be all easy going. The only trouble was that this route by-passed Narbonne altogether, to the fury of its citizens, and there can be no doubt that they were responsible for stirring up the bitter anti-Riquet campaign which followed and which made the older controversy over the route between Toulouse and the summit seem like a polite tea party.

What gave the objectors their ammunition was that Riquet's proposed high-level route would be much more costly and difficult – so difficult that they roundly declared it to be impossible. Why, it was argued, should Riquet be allowed to waste on such a crazy venture the money he wrung in taxes from the people of Languedoc when there was an easier, cheaper and more accommodating alternative. In 1673 Andreossy made a survey of the two routes and produced comparative estimates in September. These showed that while the new route would be nearly 10 miles shorter than the old it would involve in the region of 163,000 cubic yards more excavation, assuming that the canal was to be built to the new, wider dimensions.

Riquet stubbornly stuck to his guns and there can be no doubt that he was right, for not only did the low-level route involve two crossings of the Aude, but the Mediterranean littoral which he was anxious to avoid presents a very different appearance today, for there were then many areas of marshland that became shallow lakes in winter and had yet to be reclaimed. For example, one such was the Étang de Capestang, lying below and to the south of the little town of that name. This was drained at the end of the seventeenth century by the military engineer Antoine Niquet (1639–1724) by means of an elaborate system of drainage canals.

Over the line to be followed for the first $18\frac{1}{2}$ miles (30 km) eastward from Trèbes there appears to have been no serious dispute. It kept to the north side of the Aude valley, dropping by a series of locks through or near the villages of Marseillette, Puicheric, Laredorte and Homps. It was from the point where the route crossed the little river Ognon on the level that the real trouble began. For from the confluence of the Ognon, the Aude swings to the south and passes through a narrow defile between the hill of Montame on the one hand and that of Mourrel Ferrat and the Rock of Pechlaurier on the other. Having passed through this gap, the river resumes its easterly course once more. The rock of Pechlaurier falls extremely steeply towards the river. Nevertheless, Riquet maintained that it would be practicable to drive his canal round its steep flank and so maintain its position on the high ground to the north of the river through the villages of Argens, Roubia, Paraza and Ventenac. Father Mourgues, a local Jesuit priest and a clever mathematician, agreed with Riquet that this could be done, but he was almost alone. The influential Chevalier de Clerville

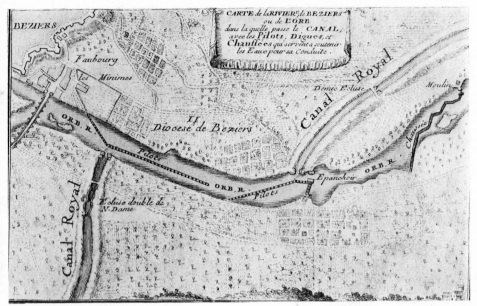

15   Map, *c.* 1697 showing Riquet's method of crossing the Orb at
       Béziers, from the Musée Paul Dupuy at Toulouse

16   Regulating weir on the Orb at Béziers

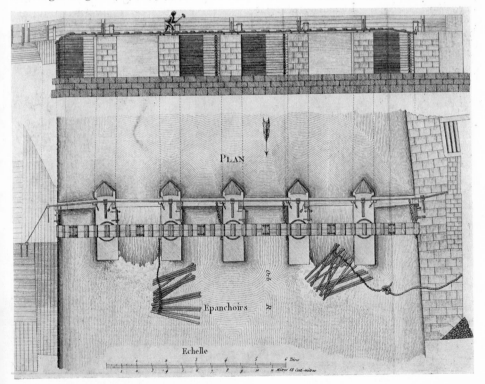

17   De Prades flood lock, river Hérault crossing

18   Round lock at Agde, showing entrance to Agde
branch cut on right

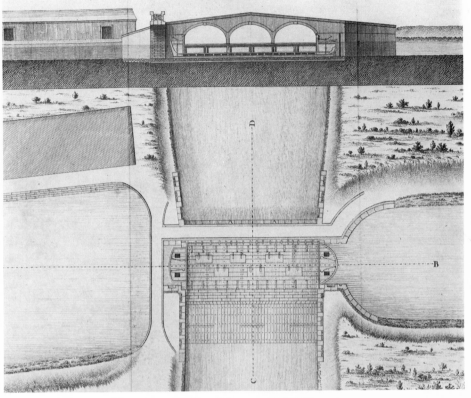

Elévation sur A B.

19 Elevation and plan of the remarkable 'Pont de Libron', showing the special Libron barge in its sunken flood position

20 Béziers: the combined lock and dry dock

21　M. Magues' splendid aqueduct of 1856 over the river Orb at Béziers

who, it will be remembered, had been responsible for checking the surveys and estimates of both sections, was among the majority who insisted that to attempt such a thing would be folly. He declared that the only practicable route through the defile was the river itself and proposed crossing the Aude at this point, a course not envisaged for the original low-level route.

Yet another low-level route was proposed by Gilade, one of Riquet's own assistant engineers who, significantly was a native of Narbonne. He proposed to serve Narbonne by joining the Aude near St Marcel, upstream of the Robine Junction, and from that junction to drive a canal in a north easterly direction across the Étang de Capestang, crossing Riquet's line east of Capestang almost at right-angles, and then by the villages of Maureilhan and Maraussan to join the Orb three miles upstream from Béziers. By keeping thus far to the north this route would avoid the hill ridge of Enserune between the villages of Poilhes and Colombiers, whose southern flank Riquet proposed to follow and ultimately to tunnel through. But Gilade's route was very circuitous and badly conceived. Indeed, one source of Riquet's strength was that his opponents were divided among themselves as to the best route to follow whereas his own intentions were perfectly clear. Having negotiated the rock at Pechlaurier, there would be one more lock to build at Argens; from the tail of this lock he would drive his canal upon one level to the right bank of the Orb directly opposite Béziers, a distance of $33\frac{1}{2}$ miles (54 km). It still remains the longest level canal pound in France.

Another trump card in Riquet's hand was that he still enjoyed the powerful support of the King and Colbert, and it was on this assurance that this stubborn, dedicated man went ahead with his 'impossible' plan. He was aware that it would need more than picks and shovels to carve a course for his canal along those rocky slopes. But Colbert wrote: 'I give orders to Berthelot to supply you all the powder you need', and it was by the aid of this that Riquet's workmen blasted a course through the rock of Pechlaurier. It must have been one of the earliest uses of gunpowder in the history of civil engineering, if not the first.

Near the village of Paraza the canal was carried over another small tributary of the Aude, the Répudre, by a single masonry arch of 30 ft span. This is the only substantial aqueduct on the whole canal which can be attributed to Riquet – the others are merely massive culverts on the Italian model. To the question why he chose to bridge this particular stream there is no convincing answer. An in-scribed stone was fixed to this aqueduct in the nineteenth century claiming (wrongly) that it is the premier 'Pont Canal' and (correctly) that it was designed by Riquet and completed in 1676. A transverse cross-section of this aqueduct reproduced in Général Andreossy's book reveals that its trough consisted of two layers of well-jointed masonry with a thick layer of some other material sand-wiched between them. What this other material was he does not tell us, but I suspect it was not rubble filling but im-pacted clay, following the method used to seal the water-slope of the St Ferreol dam.

Five and a half miles (9 km) beyond the Répudre aqueduct, Riquet carried his canal across the larger Cesse on the level by building a weir on the downstream side of that river as he had done in the case of the Fresquel and the Orbiel. One can only speculate why he did not also design a similar structure for the Cesse crossing.

East of the Cesse, Riquet's canal affords a classic example of contour canal cutting, winding this way and that, passing directly above the town of Capestang and then bending southwards to the village of Poilhes and the southern flank of the Enserune hills. In spite of following the contours, the going was slow and difficult, for much of the ground was rocky and this was no narrow canal.

While a horde of navvies, aided by blasting powder were slowly driving this long pound, Riquet's *grand bief*, eastwards towards the river Orb and Béziers, other gangs were already at work on the section from the east bank of the Orb to Agde and the Hérault as well as on the quays and breakwaters of the new port of Sète. Unlike the long pound, the section between the Orb and Hérault rivers occasioned little difficulty. No need for devious windings here; the canal was cut straight across a level coastal plain, approaching, and then running closely parallel to, the shoreline of the Mediterranean. In the 16 miles (26 km) of canal between the two rivers there are only four descending locks and only two features of interest. One is the crossing on the level of the little river Libron near the village of Vias, of which more later, and the other is the celebrated round lock at Agde which Riquet may not have lived to see completed.

The Hérault at Agde is a large and powerful river but, in order to ensure sufficient draft for traffic in dry seasons, a weir had to be built downstream of the canal crossing. This weir cut off communication between the canal and the old port of Agde lower down stream and the purpose of the round lock was to provide an economical solution to this difficulty. Through traffic on the canal passes through gates set in the round chamber at 6 o'clock and 12 o'clock. For such traffic the fall may be in either direction depending on the height of the Hérault and for this reason two pairs of gates, opening in opposite directions, are provided at 12 o'clock. Boats destined for the port of Agde, or thence to the Mediterranean, must swing through 90° in the chamber and leave via another pair of gates set at 3 o'clock. These give access to a short cut which joins the river just above Agde bridge. The fall in this direction averages 5 ft, this being the height of the weir, and to avoid having to lower the large round chamber to this extent, Agde originally worked as a double lock in this direction, boats passing into a second chamber of Riquet's normal oval pattern before being lowered to the level of the Agde cut. The bottom gates of this second chamber have long since disappeared and in any case it could not now be used in the manner intended because a bridge which has since been built across its upper end allows insufficient headroom. But the masonry walls of the second chamber still stand and Agde lock remains an outstanding example of the boundless ingenuity of Riquet and his engineers.

As the months and years slid away and Riquet's engineers and navvies were still struggling desperately to complete the long pound, his opponents intensified their attacks, for it appeared to them that Riquet was squandering their money for his own glorification on an impossibly ambitious venture that was inevitably doomed to failure. So successfully did they vilify Riquet that even Colbert began to have doubts, while in the States of Languedoc almost the only influential man who remained absolutely loyal and confident of his success was M. d'Aguesseau who, in 1673, had succeeded M. de Bezons as King's commissioner in Languedoc and inspector of works on behalf of the Province and the State Treasury.

Matters came to a head early in 1677 as Riquet's men were approaching that point along the southern slopes of the Enserune ridge where Riquet proposed to tunnel through it from south to north. It would not be a long tunnel, but it would have to be driven through sandy, crumbling stone. What was more, it was attempting something quite without precedent. No one had ever tried to drive a tunnel for a navigable canal before. Riquet's enemies noised it abroad that he was about to undertake his final *folie de grandeur*, a work where his whole misconceived enterprise would surely founder. And furthermore, they argued, even if he could achieve this bad passage (*mal pas*), on the level that he was proceeding his long pound would come to an end in an impossible situation high on a hillside overlooking the Orb and the city of Béziers beyond.

M. d'Aguesseau sent factual and impartial reports to Colbert on this situation and these, coupled with the insistent lobbying of Riquet's opponents, deeply disturbed the Minister and at last undermined his confidence and trust in Riquet. In a reply to M. d'Aguesseau, dated from St Germain, 18 February 1677, Colbert wrote:

I have just examined with care all that I have had from you on the subject of Riquet and the works on the communicating canal between Cette and the two seas. . . And to tell you in a few words of my feelings on the matter. . . I must say that it very much disturbed me because I see that, following the openings I made for you by my proceedings, you have penetrated more deeply than you were able to do the conduct of M. Riquet and the source of his views on the matter of the vast range which he has given his imagination. Although it might be best to treat him as ill, we must, nevertheless, apply ourselves with care in order that the course and strength of his imaginings does not bring on us a final and grievous end of all his works. Which is to say that he still does not produce any great result and that his works do not advance and are not completed as he would wish.

This man does as do great liars who, after telling a story three or four times, persuade themselves it is true. It has been said to him so many times, even in my own presence, that he is the inventor of this great work that in the end he has believed he is in fact the absolute author. And, on the greatness of the work he has founded [the idea of] the grandeur of the service he renders to the state and the greatness of his fortune. It is for this that he has bought a land which carries the title of baron of the States, which has made his son a Master of Requests, and which has given to his spirit, regarding the establishment of his children, a vast career and an inflation which hasn't any proportion or relationship with what he is or with what he has done.

It should be explained here that Colbert's imputation that Riquet had been purchasing honours and sinecures for himself is scarcely just. Riquet had purchased the estate of Bonrepos long before his connection with the canal began, and his title of Baron de Bonrepos was later confirmed by the Council of State. As for his eldest son, Jean Mathias, in 1673, when Riquet was seriously ill and feared he might die, he had sent him to Colbert at St Germain with a touching letter of introduction which read: 'I send you my son, as there is nobody else in the world in whom I might rightly have more confidence than him.' It was due to Colbert's influence that Jean Mathias Riquet was appointed a Master of Requests. But it is clear that Colbert had at last grown sick of Riquet's endless garrulous letters and requests for more money; sick of a canal which after eleven years was still unfinished and was surrounded by more bitter controversy than ever before. It was in this mood that Colbert wrote in terms which he must have regretted later. He went on to counsel M. d'Aguesseau to watch the conduct of Riquet and his engineers closely and to be on the look out for any extravagance or misapplication of funds; to see that men and materials were where they ought to be and to ensure that the workmen were shifting the right quantities of spoil each day and laying the right quantities of stone at the locks. The long letter ended with the threat that if M. d'Aguesseau and his assistant M. de la Feuille could not oversee the work effectively he intended to send down trustworthy men who would. Expenditure in the future must be strictly controlled and would be paid out only in strict proportion to the work actually accomplished.

This letter placed d'Aguesseau, who was loyal to Riquet, who admired what he was doing and appreciated his difficulties, in an impossible position. Rather than dissemble and play a double role, he decided to show Riquet Colbert's letter. Riquet was mortally hurt; it seemed to him inconceivable that his most powerful ally, a man who had supported his efforts so generously from the beginning, could have written such a letter. He felt that Colbert had yielded to the opposition and so betrayed him.

However, a letter from Colbert to Riquet warning him that a Commission was shortly to inspect the site of his tunnel works and that, if they felt dissatisfied they were empowered to order a suspension of further work, roused the old man's fighting spirit. He promptly reinforced the gang who were starting work on the tunnel by withdrawing all the men who were engaged on the Béziers–Agde section. Having thus mustered a huge labour force at Enserune, he deputed one of his best assistant engineers, Pascal de Nissau, to lead them into action. As fast as the miners advanced, carpenters erected massive beams to support the roof, while masons lined the side walls with stone. According to legend, in this way the first canal tunnel in the world was completed in only six days so that when the Commissioners arrived to inspect the 'impossible' work they were utterly confounded to be led through the excavation by a triumphant Riquet accompanied by workmen bearing torches.

Although this makes a very good story, it is difficult to credit. Although the Malpas tunnel is short, being a mere 180 yards (165 m) long, it is 24 ft wide and the crown of the arch is 19 ft above water-level – princely dimensions compared with English canal tunnels. Since they are far more generous than those of the bridges on the canal, it is difficult to understand why the tunnel was carved upon so grand a scale. This is not the only mystery about Malpas. Général Andreossy, in his book, shows a sectional elevation of the tunnel with two vertical shafts, one near the south end and the other near the middle. There is no sign of such shafts today and, in so short a tunnel, they would be quite superfluous either as temporary working shafts or for permanent ventilation. Admittedly the existence of these shafts would have meant that the tunnel could have been driven much more rapidly from six working faces instead of two. But this assumes that the shafts were there before the Commissioners threatened their visit, for the time lost in sinking the shafts would surely have offset any tunnelling advantage, as Riquet would have realised. Yet another curious feature of the tunnel is that, despite the obviously soft and friable nature of the rock, its western end remains to this day unlined and without a portal, so that it resembles the entrance to some large cavern. The eastern end, on the other hand, is lined, the entrance consisting of an imposing stone portal. It is tempting to think that this is evidence of Riquet's frenzied haste but, according to another theory, as built by Riquet the tunnel was never arched with stone but retained a timber lining which was later partially replaced by stone vaulting.

Riquet's triumph at Malpas was a pyrrhic victory. Soon after it he was refused any further financial support. After the long years of worry over money, the thing he had most feared had happened at last and at the very moment when ultimate success was in sight. Having been confounded by his successful tunnelling feat, his enemies prophesied that the final obstacle of the Orb valley would surely prove his downfall. But the obstinate old man (Riquet was now seventy-three) was determined to prove that he was right and the world wrong, come what might. If the world would not pay, he would, and he instructed a solicitor in Frontignan to sell all the properties which he had inherited from his family. So, financed by Riquet the work went on. 'It is said in the world,' remarked Riquet bitterly, 'that I have made a canal in order to drown myself and my family.'

The last 5 miles (8 km) of the long pound between Malpas tunnel and Béziers is markedly narrower than the rest, so much so that in one section of it one-way traffic is now enforced and passing bays provided. As there is no obvious physical reason for this, it may have been due to the new financial stringency. Spending other people's money is one thing; spending your own is quite another. But if this surmise is correct, Riquet did not allow financial considerations to sway him when the canal finally reached the rim of the Orb valley opposite his birthplace of Béziers. Once more his critics hoped for disaster and once more they were disappointed. Knowing that the long pound behind him

would provide ample water, he carried his canal down the
steep slope to the bank of the river Orb by the splendid
eight-lock staircase of Fonserannes. It was this magnificently
theatrical last gesture which finally silenced his critics. But
although Riquet may have become obsessed with his canal
and suffered from illusions of grandeur, on practical matters
he could still think coolly and rationally. The eight chambers
of Fonserannes overcome a total difference in level of
64 ft 5 in, compared with the figure of 65 ft for Cosnier's
six-chamber staircase at Rogny on the Canal de Briare.
Riquet had never forgotten that early lesson of the collapsing
lock chamber.

With the great Fonserannes staircase completed, even
Riquet's bitterest enemies were forced to acknowledge that
he had overcome his last great obstacle and that the canal
du Midi was within sight of completion at last. Colbert
wrote conciliatory letters of congratulation. Only the last
section of the canal from the river Hérault to the Étang de
Thau at les Onglous remained to be dug, with a single lock
to be built at Bagnas, before the new water route between
the Atlantic and the Mediterranean was complete. But
Riquet's last successful battle had been won at Fonserannes
and he did not live to savour final victory.

Riquet must have possessed a truly remarkable con-
stitution. Not many men of his age could have withstood so
well the fifteen years of struggle against every kind of
obstacle, physical and man-made. One feels that only his
iron determination to succeed can have kept him alive. But
now, at seventy-six, white-haired and white-bearded, the
struggle proved too much for him. On 7 September 1680,
d'Aguesseau reported to Colbert that Riquet was gravely
ill. On 1 October he is said to have summoned his son,

Jean Mathias, to his bedside and asked 'Where is the canal?'
'Just one league from the Étang de Thau,' Jean replied.
'One league,' repeated his father sadly. They were the last
words he ever spoke. So perished this indomitable man. His
bones now lie in the cathedral of St Etienne at Toulouse
beneath a black marble tablet which reads:

> En avant de ce pilier
> Sont ensevelis les restes mortels
> De Pierre Paul de RIQUET
> Baron de Bonrepos
> Auteur du Canal des deux mers
> Né à BEZIERS en 1604
> Et mort le 1er Octobre 1680.

Like the Duke of Bridgewater after him, Pierre Paul
Riquet staked his entire fortune on the success of his canal
project but, unlike the Duke, he did not live to see the
success of his enterprise. When he died he was over 2 million
livres in debt and it was his descendants who reaped the
reward. In another respect, Riquet was much less fortunate
than the Duke in that he was not his own master but had to
depend on state support. This was the prime source of the
worries that dogged him to the end. It was inevitable. In
the seventeenth century no single individual, or group of
individuals, could have raised the capital to finance the
building of a project so vast as the Canal du Midi.

The canal cost 15,249,399 livres, or the equivalent of
about 50 million gold francs. What such a sum represents in
modern currency does not bear thinking about. This
expenditure exceeded the original estimates by 6,057,399
livres and it was made up as follows: 7,484,051 from the
King's Treasury, 5,807,631 from the Province of Languedoc

and 1,957,571 from Riquet himself. These figures exclude the money expended in building the new port of Sète.

On 2 May 1681, Riquet's two sons, accompanied by M. de Lombrail, Paymaster of France, the Baron de Lanta, M. d'Aguesseau and three of the canal engineers, Andreossy, Gilade and Contigny, inspected the St Ferreol dam and the feeders in the Montagne Noire. They then proceeded to the lock at the eastern end of the Naurouze summit level which was then called Médecin but is now known as L'Écluse de la Méditerranée. Here d'Aguesseau gave the order for the sluices to be raised, the waters of the Montagne Noire thundered out, and, one after another, the pounds on the long descent to Carcassonne began to fill.

The canal was officially opened on 15 May 1681. After the Abbot of St Sernin had ceremonially blessed the waters, Cardinal de Bonzy, Archbishop of Narbonne, the elderly Anglure de Boulement and his successor as Archbishop of Toulouse, the King's Commissioners, representatives of the Provincial Parliament and other notables boarded a gaily decorated boat at the Port de l'Embouchure. This carried a band of musicians and was accompanied by a sister vessel to provide extra accommodation and to carry baggage and victuals for the journey. As a salute of cannon crashed out, as the musicians played and everyone cheered, the two boats moved off, to be followed by a procession of twenty-three barques from the Garonne loaded with merchandise. With so many boats to lock through, it must have been a slow progress, but on the 18th the procession sailed triumphantly into the Grand Bassin at Castelnaudary where a second ceremony of blessing the waters was performed, this time by the Bishop of St Papoul. As the procession went on its slow way, peasants lined the towpath shouting 'Vive le

Roi!' or 'Vive Riquet!' On the 24th they sailed through the Malpas tunnel and reached Béziers where great jollifications took place. Here the clergy left the boat, leaving the rest of the party to continue to the Étang de Thau which was reached on the 25th. The Canal du Midi was in business and it has remained in business ever since.

[FIVE]

Management and Traffic
1681–1965

For over one hundred years after Riquet's death his canal was owned and managed by his direct descendants, the line of descent and the manner in which the canal property was divided between the Riquets de Bonrepos and the Riquets de Caraman being somewhat complex and confusing.[1] Since 1673 Riquet's elder son, Jean Mathias (1638–1714), had been actively assisting his father on the canal and when his father died in 1680 he inherited a two-thirds share in the canal property and had overall responsibility for the completion of the works. Jean Mathias married three times. He had no issue from his first marriage, but two sons by his second and third marriages, Victor Pierre François and Jean Gabriel.

Riquet's younger son, Pierre Paul II (1646–1730) held a commission in the French army and appears to have played no active part in canal affairs. On his father's death he inherited a one-third share in the canal property. In 1670 he purchased the old Riquet family title of Comte de Caraman which had become extinct, rather a confusing transaction this because the Riquets de Caraman had been regarded as the senior branch of the family. This Pierre Paul II had no issue, and on his death in 1730 he bequeathed

1. See Appendix D.

his Caraman title to his nephew, Victor Pierre who, as the elder son, had inherited his father's two-thirds share in the canal. Meanwhile, his younger brother, Jean Gabriel inherited the remaining third share and became known as Baron de Bonrepos. Because this Jean Gabriel had no male issue, on his death the Bonrepos estate passed into the female line. So it was successive male representatives of the Caraman line who not only held the major share of the canal property but were also wholly responsible for the management of the canal down to the time of the French Revolution. They were: Victor Maurice, Comte de Caraman (1727–1807) and Victor Louis Charles, Comte, Marquis and finally Duc de Caraman (1762–1839), respectively great- and great-great grandsons of Pierre Paul I. The last named is regarded as the founder of the present French Caraman line. His mother was Anne d'Alsace d'Hénin Liétard, Princesse de Chimay, and his younger brother, François Joseph Philippe, Prince de Chimay, similarly founded the Belgian Caraman-Chimay family.

On the death of Riquet his two sons found themselves in an extremely difficult situation. In the first place, they had to discharge a debt of 2,090,000 livres, most of which had been incurred by their father during the final stages of canal construction. Secondly, they had to face further heavy capital expenditure, not only to complete the canal but to carry out improvements on the existing line, for it soon became apparent to them that there were serious imperfections in the work which would have to be rectified before the canal could become efficient and profitable. Lastly, their title to the ownership of the canal became the subject of a protracted legal argument that dragged on for fifty years.

In order to discharge the capital debt, the brothers were forced to sell shares of their ownership of the canal. In 1683 a certain M. Pennautier purchased one-third of the owner-ship and in 1690–1 the engineer Antoine Niquet acquired one-tenth. Altogether seven-twelfths of the estate were alienated in this way, but the Riquets wisely retained an option to repurchase and ultimately the family regained the sole ownership.

On 28 March 1683 M. d'Aguesseau began an official inspection of the canal works on behalf of the King and the States of Languedoc. After inspecting the new harbour works at Sète, he sailed across the Étang de Thau to Les Onglous in a shallop. Here he and other officials embarked in the boat *L'Heureuse* and, followed by two other craft carrying baggage and food supplies, the party set off on a voyage through the canal to the accompaniment of a salute of cannon from Marseillan. They reached Toulouse in seventy-five hours' travelling time and, as one member of the inspection party zealously recorded, it took them a total of 521 minutes to rise to the summit level at Naurouze, the equivalent of 41 seconds per foot rise. This was considered a remarkable achievement.

On 19 November, following this tour of inspection, the King's Council formally recommended acceptance of the works and this acceptance was ratified by an Order dated 16 March 1685. Nevertheless, despite this official blessing it had become increasingly obvious to the Riquet brothers that, before traffic could move freely without risk of interruption, many costly improvement works would have to be put in hand, particularly regarding the river crossings. Despite their financial difficulties, they courageously called in Sébastien Vauban to advise them and it was under his

22, 23 Fonserannes Staircase: *top*, the junction of the old (left) and new canal lines at the seventh chamber; *below*, looking up the staircase from the same position as the above

24 The oldest canal tunnel in the world: Malpas, the
western entrance

25 Map of the vicinity of Malpas Tunnel and the
Étang de Montady

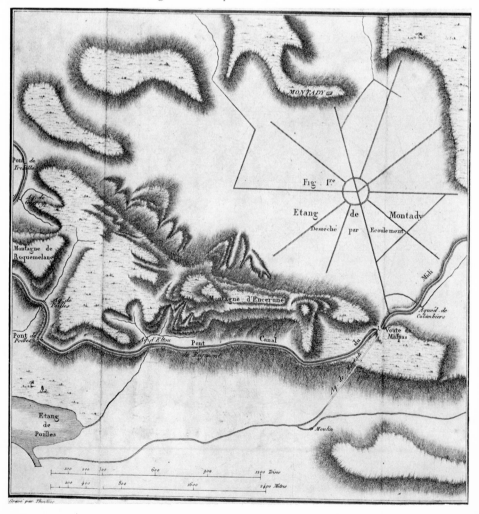

26 M. Garripuy's double 'Epanchoir à Siphon', or siphon sluice at Ventenac

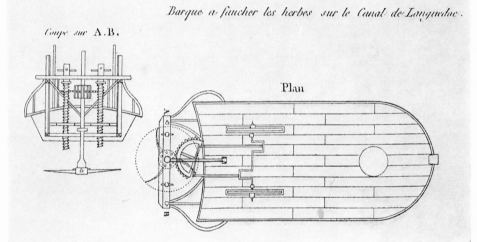

27 M. Garripuy's weed-cutting boat for the long pound (from De Pommeuse)

28   Contour Canal: the Canal du Midi and its
accompanying plane trees winds across the
landscape of southern France

29   The Cesse Aqueduct, designed by Vauban, *c.* 1686

direction that the work was carried out. When M. Toures, the King's Commissioner, inspected the canal in 1727 he congratulated the Riquets on what they had achieved. By this date it was estimated that they had spent nearly 3 million livres on improvements.

It was the death of Louis XIV in 1715 that sparked off the legal argument. There were those who held that the canal should have remained in state ownership when it could either have been managed directly by the state or by the Riquet family under the terms of a lease. Instead, the canal and its revenue had been granted as a fief to Riquet and his heirs in perpetuity, the state reserving only the right to fix toll charges and certain rights of supervision to ensure efficient management and maintenance. There can be no doubt that this curious form of agreement, which combined feudal rights with commercial management, was reached between the King and Colbert on the one hand and Riquet on the other following the precedent of the Canal de Briare where the Boutheroue family had successfully taken over and managed the works following the failure of the state to complete the canal. But the Canal du Midi was of a different order of magnitude; huge sums had been expended upon it by the King and by the provincial parliament. When the King died it was argued that any gifts or grants made by that sovereign could not be perpetual but must die with him. To this the Riquet brothers replied that at the time the grant was made in 1666 the canal did not exist and that nothing could be considered *domanial* (crown estate) before it existed. This ingenious contention appears to have completely baffled the legal fraternity who spent the next fifty years arguing the point.

Why the canal should have become the object of such a

protracted dispute is difficult to understand for it was not as though the Riquet family were sitting on a gold mine. It was not until 1724 that they began to receive any income from the canal. In 1768, by which time it had become reasonably profitable, the States of Languedoc offered to purchase the canal and for the purpose of this sale it was then valued at 8,500,000 livres. This meant that after 100 years the estimated value of the canal had only appreciated by one-third, whereas over the same period of time land values had trebled. So the Riquets certainly inherited no sinecure and it is a tribute to their stewardship that although the proposed purchase of 1768 was sanctioned by the King it was so violently opposed by the merchants and traders using the canal that they forced the provincial government to think again and the Canal du Midi remained in the hands of the Riquets.

Writing in 1822 when the long reign of the Riquets was over, de Pommeuse was full of praise for the beneficence and efficiency of what he termed 'the old management' and considered that it proved the superiority of private over public ownership. He particularly admired the resolute way the early defects in the canal had been tackled and remedied.

These troubles [he writes] could have had incalculable results due to the stoppage of navigation and losses in receipts had it not been for the activity displayed in carrying out repairs, the incessant inspection and interested zeal of the owners. That is a powerful point in favour of private management, and one cannot conceal the fact that in similar circumstances, the unavoidable loss of time and the kind of responsibility inherent in public administration would have left plenty of time for these mishaps to worsen and be without any remedy.

He goes on to quote one of Riquet's descendant's own statements of policy:

In undertaking the work, we insisted not upon securing the lowest price, but in having the work done as perfectly as possible. A constant attention to maintenance, sometimes averted serious disasters, but if they did happen, the best policy, in our eyes, was to pay more so as to obtain greater speed and better workmanship. And it was not enough to repair unexpected failures, it was also worthwhile to prevent those likely to happen. Public administrators would have been accused of chimerical forethought in asking for money for uncertain dangers, whereas we were driven by honour and interest frequently to spend considerable sums to prevent a calamity we thought probable, albeit as a distant date.[1]

The old management consisted of a head office at Toulouse staffed by a general manager, a general controller and a chief toll collector, these three forming a council whose duty it was to render regular reports to the owners on all canal matters, to receive instructions from them and to pass these on as necessary to the district staff. There was also an archivist at Toulouse in charge of a depository where all papers relating to the construction, maintenance and management of the canal were preserved. The practice of employing an archivist has continued down the years with

1. De Pommeuse does not quote his sources, but in fact this quotation comes almost verbatim from the *Histoire du Canal du Languedoc* (Paris: Imprimerie de Crapelet, 1805). This book was written anonymously by the 'Héritiers de Riquet'. Since the Riquets de Caraman were in exile at the time it was written, its author is presumed to have been Mathias Riquet de Bonrepos who apparently took the leading part in the administration of the canal at this period. Because Jean Gabriel, Baron de Bonrepos, died without male issue (see Appendix D), the precise relationship of this Mathias is obscure.

the result that no transport organisation in the world today has a more complete set of records and certainly not one extending over so long a period of time.

Under the general manager there were seven district managers based on Toulouse, St Ferreol, Castelnaudary, Trèbes, Le Somail, Béziers and Agde. At each of these places there was also a toll collector, one controller and one 'visitor'. What was the precise role of the last-named is not clear. He may have been a travelling inspector, for we are told that he and the controller acted as supervisors and overseers whenever any repair work was going on in their district during stoppages. Under this management there were one hundred lock keepers and eighteen guards. The latter were probably what we should call lengthmen with special responsibilities for certain flood sluices or flood gates. For it is later recorded that one guard was appointed to take charge of the peculiar 'floating aqueduct' across the Libron which will be described in the next chapter.

Except in cases of emergency, the canal was only stopped for repairs once annually during the summer months when all necessary work was carried out to a carefully prearranged plan.[1] Each year each district manager sent to the Toulouse office a report of the work that needed to be done in his division and from these reports the general manager

1. The *Histoire du Canal du Languedoc* contains the almost unbelievable information that, at the outset, the canal was closed to traffic for periods ranging between six weeks and two months every six months. One suspects that this must have held good only during the period when the improvement works were being carried out by Vauban and Niquet. According to the same authority, major stoppages later took place at two-yearly intervals. Apparently on these occasions the canal was drained throughout. It took four days to empty and six days to refill.

prepared an overall schedule for submission to the owners. This schedule was checked by the owners when, accompanied by the general manager, they made their annual tour of inspection in what de Pommeuse describes as 'a kind of gondola'. For the works thus approved, contract drawings, specifications and instructions were prepared at Toulouse and sent out to the district managers concerned who were then responsible for detailing them and for engaging local contractors.

Judging from this extract from the management instructions which de Pommeuse quotes, the canal was run on paternalistic lines:

Every employee who behaves well must be a friend of the owners; his needs, his worries and his children must be your concern; you must spare nothing to soothe their minds because duties of such importance cannot be done by people distracted with worries.

The owners must strive to keep their people in that quiet mood that clears every faculty of the mind. Your kindness and your care to fathom their wants by overcoming that shyness which stops people from voicing their complaints, are means to capture honest hearts and keep employees as much interested as the owners in the maintenance of this great and important piece of work.[1]

Such an attitude towards the canal employees was singularly enlightened in the eighteenth century and it certainly paid dividends. It laid the foundations of a great feeling of *esprit de corps* throughout the length of the canal which, because it was primarily a sense of loyalty to the Canal du Midi itself rather than to any particular owner of it,

1. The source of this quotation, like the previous one, is the *Histoire du Canal du Languedoc* (1805).

proved strong enough to survive all vicissitudes down to the present day.

Needless to say, it was the outbreak of the French Revolution in 1789 that set a term to the 125 years of exclusive ownership of the Canal du Midi by the Riquet family. The Bonrepos branch remained in France and somehow contrived to retain their estates including their third share, but the loyalist Caraman branch emigrated in 1792 and their estates, including their two-thirds share in the canal, were confiscated by the Republican Government. Under the Emperor Napoleon, the Compagnie Général du Canal du Midi was set up by a decree dated 10 March 1810, and to this the Caraman holding in the canal was transferred. The new company had a nominal capital of 10 million francs divided into 1000 shares, out of which Napoleon appropriated 900, distributing them among his favourite marshals and such other members of his newly created aristocracy as he wished to reward. The general meetings of this company were presided over by the Grand Chancellor of the Legion of Honour, but the famous Corps des Ingénieurs des Ponts et Chaussées[1] was made responsible for the day-to-day running and maintenance of the canal.

This was the set-up at the time de Pommeuse visited the canal in 1818. He was able, through the courtesy of M. Clauzade, the then Chief Engineer, to travel through it in the latter's inspection gondola. Evidently the upheaval of

---

1. This celebrated Corps of civil engineers was established in 1719. It grew naturally out of a Corps des Ingénieurs de Génie Militaire which had been founded at the suggestion of Vauban towards the end of the seventeenth century. In 1747 the equally famous École des Ponts et Chaussées was founded under the direction of J. R. Perronet to train new recruits for the service.

the Revolution had done little to change things on the canal, for he records how pleased he was to discover, all along the route, 'that hereditary, almost family zeal of the canal employees'. The form of the administration was little changed either, except for a significant new emphasis on engineers. Whereas before one can only infer that the 'managers' possessed engineering experience, now there was a chief engineer at Toulouse with a staff of five district engineers and three sub-engineers.

Another new post to appear was that of plantation manager, and we are told that he had charge of four nurseries at Toulouse, Naurouze, Trèbes and le Somail where trees were raised from seed for planting out on the canal banks. There were in these nurseries over 150,000 young trees and 325,000 seedlings. Varieties mentioned include oak, ash, Italian poplar, larch and other 'resinous trees' but not, curiously enough, the plane tree which was later to surplant all others on the banks of the canal, perhaps because its roots were found to hold the banks better than other species.

After the Restoration, a Royal Ordinance of 25 April 1823 decreed that the shares in the Canal Company distributed by Napoleon should revert to the Riquet family, the Caraman branch receiving 292 newly created shares by way of compensation. The Canal Company was at this time thriving but, as in England, the coming of the railway appears to have thrown its proprietors into a panic which caused them to throw up the sponge without a struggle. Construction of the Chemin de Fer du Midi began in 1851 and the section between Bordeaux and Toulouse was opened in August 1856. The section from Toulouse through Narbonne to Sète was opened on 22 April 1857 and little over a year later the Canal Company had agreed to lease their canal to the railway company. This lease was signed on 28 May 1858 and ratified by the state in the following month. Very fortunately for the canal, however, this was not a lease in perpetuity but was dated to expire in June 1898.

When this expiry date became imminent, the canal proprietors agreed in November 1896 to transfer the Canal du Midi and its extension, the Canal Latéral à la Garonne, to the state, an agreement which was ratified and authorised by an Act of 27 November 1897. So the canal passed directly out of railway control into state management and, as on all state-owned French waterways, tolls were abolished under that enlightened principle which regards all inland waterways as valuable national amenities and, as such, not expected to 'pay their way'.[1] As a result, traffic on the canal slowly began to recover and, with the exception of the second world war period, this trend has continued.

When acceptance of the canal works was formally confirmed in March 1685 a second Order fixing the toll rates was issued and a special private court was set up in Toulouse for the purpose of arbitrating in the event of any dispute between traders or local landowners and the management concerning the operation or maintenance of the canal. In addition, inspectors appointed by the King and by the

1. This applied throughout France until 9 April 1953. Since that date each commercial craft owner pays for the use of the inland waterways in two ways. First, an annual 'business tax' of approximately 260 fr per boat which is paid to the central exchequer. Secondly, a nominal toll of 20 centimes per ton/kilometre which is estimated to cover 25 per cent of the cost of maintenance. In the case of visiting pleasure boats these charges are not levied. Hire cruisers using a canal regularly have to pay the annual 'business tax'.

regional parliament in Languedoc continued to exercise their right of supervision.

For toll purposes the canal was divided into 48 sections, each approximately 5 km long. Tolls were at the rate of 4 centimes per kg-quintal immediately after the Revolution, although this rate was reduced in the case of building materials, coal, sand, gravel and manure. There were no mooring dues at the canal ports and barges travelling light went toll-free.

Of the traffic that used the canal in its earliest days very little appears to be known. One thing is certain, however, and that is that it was never used by shipping as an alternative to the much longer sea route via the Straits of Gibraltar as was envisaged at the time a canal between the two seas was first conceived. Much of the goods that would have been routed by sea may have been diverted to the canal, but it was not carried in the same bottoms. This was particularly true of the period of the Napoleonic war (1800–15) when traffic on the Canal du Midi increased markedly as a result of the British naval blockade. But the only recorded use of the waterway as a ship canal was somewhat bizarre and occurred in 1690. A naval victory over the combined English and Dutch fleets at Tourville went to the head of Louis XIV and encouraged him to plan the invasion of England. To this end, a number of what are described as 'rowing galleys' were despatched through the canal from the Mediterranean to the Atlantic, their ultimate target being Teignmouth on the coast of South Devon. Like Napoleon and Hitler, however, Louis had second thoughts and his invasion fleet never arrived at its destination.

The main reason why the Canal du Midi never became a commercial through route for shipping was the state of the river Garonne. From the days of Pierre Paul Riquet to the present time the Garonne has always been an open river, that is to say without locks, from its source in the Val d'Aran in the Spanish Pyrenees until it meets the Dordogne and the combined rivers form the Gironde estuary. Because it is a torrential river, subject to violent fluctuations of flow, and because there is considerable fall between Toulouse and Bordeaux, navigation between the two cities was an extremely hazardous exercise. It could only be carried on by small craft piloted by skilful local men with great knowledge of the river. Even so, trade was frequently interrupted either by floods or by lack of water. Looking at the river today it is difficult to conceive how it was ever navigated at all by craft of commercial burthen. Goods were frequently transhipped from canal barges into smaller river boats in the Port de l'Embouchure at Toulouse. Even so, it required as many as twelve horses to haul one of these river craft upstream from Bordeaux to Toulouse. Whenever a boat ran aground on a shoal, the crew would go over the side and, wading in the water, unload their cargo onto the bank until the boat was light enough to be manhandled over the shallows and then reloaded. Despite such time-consuming and laborious expedients, there were many occasions during the summer months when the river level fell so low as to be impassable.

These were the conditions which led to early proposals for a canal to by-pass this difficult river, but these were not fulfilled until 1856 when the Canal Latéral à la Garonne was opened throughout from Toulouse to a junction with the tidal river at Castets-en-Dorthe, 35 miles (56 km) above Bordeaux. Unfortunately, this date almost coincided with that of the leasing of the entire route to the competing railway company. It is surprising that, despite railway competition and the new Lateral Canal, traffic on the river should have continued into the latter part of the nineteenth century. Indeed, there are old retired boatmen still living who can remember navigating the river as young men. There were two reasons for this persistence; first, the avoidance of tolls on the canal and, secondly, the fact that when river conditions were favourable the downstream journey was made far more rapidly than by the new canal. Intermediate navigable connections were provided between river and canal at Montauban, Moissac, Agen and Buzet. As late as 1885, traffic on the Garonne between Toulouse and Bordeaux amounted to 6,032,817 ton/km, but it is significant that 95 per cent of this total was carried downstream. When the canal was made toll-free in 1898 this river traffic soon ceased.

Over twenty years before work on the Garonne Lateral Canal began, the eastern end of the Canal du Midi had been provided with an inland link with the Rhône and so with the whole central system of inland waterways. Before Riquet had completed his canal, the States of Languedoc had commenced cutting the first part of this canal to the Rhône. The work consisted of enlarging ancient medieval channels through and between the shallow *étangs* and salt marshes from Sète to the ancient walled town of Aigues-Mortes and when completed this route was known as the Canal des Étangs. If it is true that Riquet envisaged that his canal would soon be linked by inland waterways with the Rhône he was unduly optimistic, for over a hundred years were to pass before the canal at Aigues-Mortes was united with the Rhône at Beaucaire by an extension cut across the Camargue through the old stranded port of St Gilles.

The history of this Canal de Beaucaire makes a sorry contrast to that of the Canal du Midi. A concession to construct it similar to that granted to Riquet was given to the Marshal de Noailles but, unlike the Riquets, the Marshal and his family did precisely nothing for nearly fifty years, after which the concession was revoked. It was then granted to a company formed by Marshal de Richelieu, but once again nothing was achieved until the States of Languedoc took matters into their own hands. They then spent nineteen years haggling with local landowners over fishing rights and the draining of marshes before construction was finally begun in 1777. The canal was still incomplete when work was interrupted by the suppression of the States Parliament at the time of the Revolution. The new Government then authorised a company which finally completed the link with the Rhône in 1808. The Canal de Beaucaire and the Canal des Étangs are now together known as the Canal du Rhône à Sète.

It is doubtful whether Riquet envisaged horse haulage when he planned the Canal du Midi and it seems likely that the first boats to use the canal were small craft of coastal type with hinged masts which were hauled by men when they could not sail. We know that the craft in which the Archbishops of Toulouse and Narbonne travelled on the occasion of the ceremonial opening was a barque measuring 57 ft long by 12 ft beam. By the mid-eighteenth century sailing craft were still in use but were horse-hauled: 150 of these boats were said to be trading regularly on the canal. They loaded from 88 to 98 tons, were manned by a master and one sailor and were drawn by one or two horses in charge of a postillion.

By the time de Pommeuse visited the canal in 1818, traffic

had increased for he tells us that at several points the number of boats passing annually over the previous eighteen years had averaged 1920, carrying a total of 75,000 tons. There had been a change by this time from the old small, coastal type of craft, for these boats, though still decked, are described as having flat bottoms and loading from 100 to 120 tons on a maximum draught 1·6 metres or approximately 5 ft 4 in, although this had to be reduced to 1 metre in the case of boats trading down the Robine to Port la Nouvelle. These barges (for such they undoubtedly were) displayed graduated displacement scales so that the toll clerks could readily estimate the tonnage dues payable. Some of them, at least, still carried sail at this date for de Pommeuse writes: 'They are driven by a master with one assistant, sail propelled or towed by one or two horses led by a coachman.'

The annual budget of the canal averaged out over the same eighteen year period worked out as follows:[1]

	Francs
Income from tolls, 75,000 tons at 19·28fr per ton	1,446,000
Other income (rent from land, mills etc.)	64,000
TOTAL	1,510,000
Maintenance and Administration expenses, average year	710,700
Nett income, or available for improvements	799,300

Presumably horses were now responsible for towage throughout the length of the canal and sails were carried mainly for the passage of the Étang de Thau, for the mind

1. Figures supplied to de Pommeuse by M. Clauzade, Chief Engineer. For a breakdown of maintenance and administration expenses see Appendix B.

boggles at the idea of craft sailing along Riquet's tortuous long pound. The difficulty of navigating the Étang de Thau was solved about 1834, however, when the Company's steam tugs, *Riquet*, *Colbert* and *Vauban* began to ply regularly between the mouth of the canal at Les Onglous and the port of Sète. It is safe to assume that once this service had been introduced, the Midi barges soon ceased to carry sail.

The introduction of steam tugs was accompanied by an increase in traffic and in 1838 there were reported to be 273 barges plying regularly on the canal, a figure which excluded traders in river and coastal craft from the Garonne, the Rhône and the Mediterranean. Presumably this total represents barges operated by trading companies or captains owning their own craft (or 'Number Ones' as we would call them) and does not include the fleets owned and operated by the Canal Company itself. There were two of these fleets. One consisted of so-called 'accelerated boats', the equivalent of the 'fly boats' on our canals, which were used for urgent or perishable traffic, loaded only from 50 to 60 tons and travelled at nearly twice the speed (6 k.p.h.) of the heavier barges. These fly boats also carried passengers at a fare of 15 centimes for each of the 48 five-kilometre stages covered. The second fleet consisted of packet boats carrying passengers and mails. These could accommodate 150 passengers in two classes of cabin, a 'common room' and an 'ornamental saloon', saloon passengers paying at the rate of 25 centimes per stage. The rate of travel of these packet boats was 11 k.p.h.

To ensure the speedy transit of mails (and, incidentally, a saving in water), in the early days of this service these *barques de poste*, as they were called, did not pass through the major lock staircases but shuttled over the pounds between them. But this meant that many more boats and crews were needed to maintain the service than would otherwise have been the case. Also, passengers complained bitterly at having to make repeated changes, hurrying down the tow-path with their belongings. So the service then settled down to a leisurely four days with three overnight stops. Even so, this through service did not use the Orb crossing. At noon on the fourth day, the Toulouse packet terminated at the head of the Fonserannes staircase, passengers and mail being conveyed into Béziers by road in order to catch a second boat east of the Orb for the remainder of the journey.

Apparently this leisurely service was still in operation when de Pommeuse visited the canal in 1818, for he com-pares it unfavourably with similar services on English canals. 'After what we have seen of the operation of packet boats on the English canals', he writes, 'one is entitled to be surprised at the slowness of said "barques de poste", which take four days for the whole Languedoc Canal trip and compel passengers unpleasantly to sleep at Castelnaudary, Carcassonne and le Somail.' He goes on to say that the Canal Company was currently employing 18 masters, 20 coach-men and 42 horses in the packet boat service.

De Pommeuse's criticism was a little unfair because, compared with English canals, the locks and bridges of the Midi Canal are most inconveniently laid out for horse towage, as we shall see in the next chapter. It is therefore remarkable that the packet boat schedule between Toulouse and Béziers was finally reduced to thirty-six hours, travelling night and day. The service was then made the direct respon-sibility of the chief engineer and was managed by a chief *inspecteur-receveur* with the help of a *contrôleur* and two assistants. There were seven itinerant collectors who acted

as guards cum ticket inspectors since they were responsible
for the running of the boats and for collecting fares. There
was evidently a connecting service to Narbonne, for we are
told that of the twenty-three masters or captains of the
packet boats, two were confined to the Canal de Jonction.
It must have required heroic efforts to maintain such a
service, but it could not prevail against the Chemin de Fer
du Midi, and when the canal was leased to the railway
company the packet boats soon vanished for ever.

From the beginning of this century there was a slow but
steady increase in traffic out of the trough into which it had
fallen during the term of the canal's lease to the railway
company. This recovery was undoubtedly greatly helped by
the fact that the Canal du Midi, in common with other
French waterways, was freed from tolls in 1898. A second
factor which helped recovery was the changeover from horse
haulage to diesel-powered barges which began in the early
1920s. Because of the larger tonnage of the boats used and
the inconvenience of the canal for horse-towage, motorised
barges had a much greater advantage over their horsedrawn
predecessors than was the case on the English narrow canal
system. So, whereas in England the horse-boat stubbornly
persisted, on the Canal du Midi it very rapidly disappeared.

The first boat to be motorised in 1925 was the *Ville de
Sète*, belonging to the Compagnie Général de la Navigation,
a company which operated throughout the French inland
waterway system. Her success was such that this company
speedily converted the rest of its Midi Canal fleet, the *Ville
de Bordeaux*, *Ville d'Agen*, *Ville de Béziers* and others. The
next to follow suit was the Compagnie Enérgique, a company owned by B.P. Ltd, which motorised its *Xerxes III* in
1926. Boats belonging to this and other companies were so

Le Somail, once the 'packet port' for Narbonne

Riquet's Répudre Aqueduct of 1676

32, 33   L'Écluse l'Evêque: a typical Midi Canal lock
and lock-house

quickly converted after this that by 1930 all company-owned boats had been motorised. The effect of this was that owner-boatmen with horse-drawn craft on the canal found it increasingly difficult to obtain cargoes and were forced to follow suit. As a result, the last horses disappeared from the canal in 1933/34. A long era came to an end when the mule-hauled *Margdeline* was converted in 1935.

The typical Midi horse-boats were timber built with grace-ful, curving bow, prominent stem-post and a large wooden rudder aft swung by a wooden tiller-bar of truly heroic proportions. As originally motorised, the wooden tiller was retained, but later they were converted to wheel steering in association with a small wheelhouse for weather protection. In order to clear low bridges when running empty, these wheelhouses were designed to be easily dismantled or collapsed. Plate 38 shows the *Ste Germaine*, a converted horse-boat engaged in the wine trade, still with her original wooden tiller. Incidentally, her Captain, M. Coucière, now owns a modern steel wine tanker barge, *La Belle Paule*, which is one of the best-maintained craft on the canal at the present time. Steel barges of this kind are replacing the older wooden ex horse-boats, although there are still a few of the latter to be seen.

In the period of the occupation during the second world war, no fuel was available for powered boats and horse haulage had to be temporarily reintroduced, the necessary horses being supplied by the government. The only powered boat to continue in service on the canal at this time was the *St Jacques*, which was requisitioned by the Germans and fitted with a gas producer using wood fuel. For some unknown reason, during this war period mules were used in preference to horses on the section between Carcassonne and Marseillette only. Records of the daily number of boats and tonnage passing the St Roch staircase at Castelnaudary show, not surprisingly, that during these war years traffic dwindled to an average of a single boat per day, an all-time low that was probably not plumbed even during the period of railway control for which no figures are available. It certainly averaged four boats a day in 1900 when there had been little time for recovery.

Because the canal passes through a purely agricultural region and the few large towns on its route have never been heavily industrialised, since the coming of the railway there has been little traffic to, or originating from, intermediate wharves, many of the smaller ports falling into disuse. The bulk of the hauls became long-distance, a characteristic that has continued to the present day. Though there have been fluctuations since the last war, the general trend of canal trade has been upward, the main reason for this growth being the traffic in hydro-carbon fuels conveyed in tanker barges. This has followed the establishment of oil refineries at Frontignan on the Rhône–Sète Canal and at Port la Nouvelle. The factor limiting traffic growth has been the number of barges available. The reason for this is that the Canal du Midi and the Garonne Lateral Canal can only admit craft of 145 tons and 170 tons deadweight[1] and since this is well below the accepted European minimum standard, traders hesitate to invest capital in building new barges which, because of their small size, do not compete effectively against modern developments in rail and road transport.

The following is a picture of the modern traffic pattern

1. These are the official authorised figures, but in fact barges have been known to pass through the Canal du Midi with a deadweight of 160 tons.

extracted from an article by M. Pfaff, then Chief Engineer of the canal, which appeared in a special edition of *Regards sur la France* in 1966. The figures given are for the year 1965. Since that date, traffic has fallen slightly on the Canal du Midi but the pattern of trade remains the same. On 1 January 1965 there were, plying regularly on the Midi and Lateral canals, 81 general cargo vessels, 54 tankers for the transport of hydrocarbons, 5 tankers for tar and its derivatives and 26 wine tankers of which 22 were equipped with removable container tanks. The normal average journey time of these boats is 75 hours, or from 5 to 7 days depending on the season, between Toulouse and Sète, and 60 hours or 4 to 6 days between Toulouse and Port la Nouvelle. The principal westbound traffics consist of hydrocarbons from Frontignan and la Nouvelle to oil depots in the Toulouse region; wine traffic from the Lower Languedoc to Toulouse and, especially, Bordeaux; seasonal traffic in salt from the Mediterranean salt pans to Bordeaux to supply the Atlantic fishing fleets. Eastbound traffic consists mainly of the maize which is grown extensively in the south-west. This is shipped from storage silos situated in the Toulouse region and at various depots along the Lateral Canal and is carried to Sète and la Nouvelle for export.

Through traffic to or from Bordeaux via the Lateral Canal and the river Garonne takes from $3\frac{1}{2}$ to 5 days (50 hours) between Toulouse and Bordeaux and from 4 to 6 days (55 hours) in the opposite direction, the difference being due to the powerful current on the tidal Garonne. For this reason, barges are fitted with larger engines than would be necessary if their use was confined to still waters only. Total annual traffic passing through the Canal du Midi in 1965 amounted to 277,714 tons or 50,751,544 ton/

km, the average number of boats passing daily through the
St Roch locks during this year being slightly over ten.
Assuming that de Pommeuse's traffic figures are correct, this
represents a vast increase over the tonnage carried in pre-
railway days.

In addition to the traffic it receives from the Canal du
Midi, the Lateral Canal has a considerable trade of its own
consisting of hydrocarbons from the Gironde refineries,
woodpulp, pyrites and cereals. And because the permitted
draught (1·8 metres or 6 ft) is greater than on the Midi,
barges confined to the Lateral Canal can have an average
deadweight of 170 tons. Consequently, total traffic on the
Lateral Canal amounted to 443,507 tons or over 38 million
ton/km in 1965.

Today, in addition to the carrying company fleets, a
considerable number of barges are 'Number Ones', owned
by their captains, and their immaculate condition reflects
the pride of ownership. Boats such as *La Belle Paule*, *Stella*
or *Serge Gérard* are a delight to behold. Occasionally a
captain may have a home ashore and work his barge with
the help of a paid hand, but in most cases the barge is the
family home, the children leaving her during term time to
attend boarding school in Toulouse. Because the Midi
Canal barges are small by French standards, they are usually
confined to the hauls mentioned above, but a few of the
more recently built steel craft have engines of 200 h.p. or
more and are thus able to ascend the Rhône. They are
therefore capable of picking up cargoes anywhere on the
waterway network of northern France and occasionally do
so.

It is by pleasure rather than by commercial traffic that
Riquet's canal is still used as he intended. An increasing

number of British and continental yachtsmen make use of the Canal du Midi each year as a convenient passage between the Western Atlantic and the Mediterranean cruising grounds, the more so since it is free of toll so far as pleasure traffic is concerned. In addition, a flourishing pleasure cruiser hire business has been established recently with headquarters at Castelnaudary. Pierre Paul Riquet would be gratified by the fact that, after nearly three centuries of use, there is certainly no lack of traffic on his canal today.

[SIX]

The Canal Today –
Les Onglous to Trèbes

The Étang de Thau is a lake of very considerable extent. So large is it that a stranger travelling by road and finding himself in one of the ancient little ports of Bouzigues, Mèze or Marseillan on its western shore might be forgiven for thinking that he had already reached the Mediterranean coast. For the low and narrow isthmus to the south-east, which cuts it off from the sea and carries the main road and railway line from Agde to Sète, is not easy to see on the horizon, particularly in hazy weather. From the entrance to the harbour of Marseillan, looking south-west, two stone jetties can be seen, the longer eastern one terminating in a small red-and-white painted lighthouse. This is the Phare des Onglous (two-flash red light, 6 second period, white area 210° to 229°) and it marks the mouth of the Canal du Midi.

To enter the canal at this point is a strange experience and to leave it is even odder. For it appears to the canal traveller as though he is sailing straight from the canal into the sea without passing through any intermediate basin or sea-lock. As an introduction to the Canal du Midi it is scarcely an attractive starting point, for this is mosquito country, a flat region of marshes, salt pans and shallow lakes only relieved by the single humped volcanic hill of Mt St Loup to the south. After skirting the southern shore of the little Étang de Bagnas, the canal swings north to Bagnas lock, presumably the last to be built on the canal.

As at all the original lock sites throughout the canal, the lock chamber at Bagnas is of the characteristic Riquet form and dimensions. Whether it is the original masonry or not it is impossible to say, but it is reasonable to suppose that after the lapse of nearly 300 years most of the lock chambers have been rebuilt at some time or another. De Lalande double lock, west of Carcassonne, for example, bears the date 1786 cut into one of the side wall stones. Timber lock gates of the original pattern with balance beams survived until well into the nineteenth century before, at some date which has not been recorded, they were replaced by the present design of metal-framed gates without balance beams. Large gates of this type which lack the counterpoise of a balance beam are often provided with a roller at the foot of the mitre post[1] which runs on a metal track on the sill and so supports the weight of the gate. The lock gates of the Canal du Midi, however, are supported entirely from the axis of their heel-posts,[2] the collars which hold them in the original hollow quoins[3] in the lock walls being strengthened and fitted with journal bearings, originally of brass but now of steel. Despite the considerable load on these bearings, the gates swing quite easily and are opened and closed by a long horizontal rack-rod of cast iron. A pinion attached to a vertical shaft having a fixed crank handle at its upper end engages this rack. On approaching one lock, we found that

1. The vertical member of the gate furthest from its hanging.
2. The vertical member of the gate nearest to its hanging.
3. The semicircular recess in the masonry in which the heel-post semi-rotates.

on one of the gates the rack-rod had just snapped off so we had to swing the gate by attaching a rope to the mitre-post and were surprised to find how easily it moved.

There are now two sluices in each gate and the old method of lifting them by screw and capstan bar has given place to the more orthodox rack and pinion gear. Alternative spindles are provided, one direct and the other giving a slower lift through step-down gearing. Windlasses are detachable but are not carried by the boatmen. Instead, they are available at each lock gate where they are carried in clips on the walkway railings when not in use. One reflects wryly that if this practice were to be adopted in England there would soon be no windlasses left.

No steps or landing platforms are provided at the lower end of the locks. Those wishing to leave or to rejoin a boat must needs use a vertical iron ladder on the wing walls below the lower gates. These locks must have been very inconvenient to work through with horsedrawn boats, particularly in some cases where bridges with no towpaths beneath them span the lock tails.

The attractive lock cottages with their pantiled roofs and faded green window shutters look as though they had been weathered by years of strong sunlight as, indeed, they have. In the nineteenth century, finely lettered rectangular plaques of cast iron were mounted over the doorway of each cottage. These bear the name of the lock and, below it, the distance in metres to the next lock in either direction. Lock keepers are now in telephonic communication with each other so that they receive advance warning of an approaching boat. This means that it is usual to find the locks ready unless traffic dictates otherwise. Lock keepers assist in working the locks but boat crews are expected to do their fair share. And here a note of warning: the English practice of 'dropping the paddles' by releasing the pawls from the ratchets and allowing them to run back by their own weight is strictly *interdit* on the Canal du Midi. Perhaps this is one reason why it is so rare to find a lock sluice that is out of order.

Passing under the Pont St Bauzille, the canal leaves the open marshland and is soon under the welcome shade of trees on the approach to de Prades lock. This lock, which has a rectangular chamber, either did not exist in Riquet's day or it replaces a single pair of flood gates. It is a flood lock, built to prevent the floodwaters of the river Hérault, just beyond, from invading the canal pound as far as Bagnas Lock. When the river level is normal, all the gates of de Prades stand open but, like all rivers in this area, the Hérault is subject to sudden spates. For example, when we made our spring voyage from Castelnaudary to Marseillan we found the river in spate, and the flood lock in use and were surprised to be told that had we arrived a day earlier we should have been unable to make the crossing. Thus we learned at first-hand the practical disadvantages of Riquet's method of crossing rivers on the level. However, the difficulty and delay so caused are by no means so great now as they were in the days of horse-drawn craft. In those days barges were guided across the Hérault by means of a cable whose anchor blocks may still be seen by the mouth of the canal near de Prades lock. Meanwhile the towing horses were either taken on board or else walked round via Agde bridge.

The entrance to the canal opposite is some little distance downstream and, because the bank is thickly clothed with trees and bushes, it is by no means easy to see until it opens up. Consequently the unwary or unobservant stranger may suddenly find himself rapidly approaching Agde weir on a

swift current. From the river, it is a mere 200 yards or so to
the famous *écluse ronde* at Agde. Here the branch cut leading
to the old port of Agde below the weir looks depressingly
narrow, shallow and disused – very like an English narrow
canal in fact. Yet appearances are deceptive for it was here
that, on my second voyage, we were able to enter the canal
from the Mediterranean[1] in a yacht drawing nearly 5 ft of
water. A word of warning, however. In such circumstances
if the round lock is closed, it is unwise to lie beneath the
low bridge across the tail of the lock because the level of the
Agde cut, like that of the river Hérault, is subject to sudden
fluctuations.

Beyond the little canal port of Agde, just past the round
lock, there is little to remark, the canal traversing a dreary
tract of marshland close to the coastline of the Mediterranean until, at the end of a long straight, the voyager sees
ahead of him a forbidding array of steel girders and menacing
guillotine sluice gates spanning the canal. These represent
an ingenious modern solution to the problem of the Libron
crossing, one of the most persistent trouble-spots on the
entire canal.

Under normal conditions the Libron is an insignificant
stream, but it is subject to violent flash-floods which may
occur as often as twenty times in a year. These are aggravated
when they coincide with an on-shore wind in the Mediterranean as this causes the Libron water to back-up. In the
period immediately following the opening of the canal it

1. Two jetties, each with a lighthouse, mark the entrance to the
Hérault river and the port of Agde from the sea. The lighthouse on the
western jetty has a red top giving two 6 second flashes of a red light at
night, while that on the eastern side has a green top and gives 4 second
flashes of a white light.

was said that one such flood could fill the bed of the canal with silt for a distance of half a league. Owing to the flatness of the terrain (the bed of the canal is actually below the level of the sea at this point) and the proximity of the site to the mouth of the Libron, this was a trouble which no conventional aqueduct could cure. Hence the unique 'Libron Raft' or 'Pontoon Aqueduct' was evolved. Both banks of the canal were walled up to the crossing point of the Libron. A special form of flush-decked barge was then built. This measured 100 ft long by 20 ft beam, had provision for scuttling and was fitted at each end with adjustable ramps. This curious craft was normally kept moored in a special bay beside the crossing, but whenever a spate threatened it was moved out between the retaining walls until it spanned the stream and was there sunk. The floodwaters then passed over the flush deck of the barge and between the raised ramps and were thus kept out of the bed of the canal. When the flood had subsided, the barge was refloated by pumping it out with an Archimedean screw pump and it was then returned to its mooring to await the next inundation.

Whenever this peculiar *pont de Libron* was in use it blocked the canal to traffic and the present arrangement is designed to obviate such delays. At the point of crossing, the Libron has now been provided with twin channels situated 100 ft apart. Guillotine sluice gates can be used to divert floodwater into either channel at will and, by the closure of four similar sluices, either channel can be isolated from the canal. Thus when the Libron is in spate, a barge passes beneath two open sluice gates and then stops in the central chamber between the two channels. The gates behind it are lowered so that the floodwater may be diverted and the protective gates ahead can then be raised.

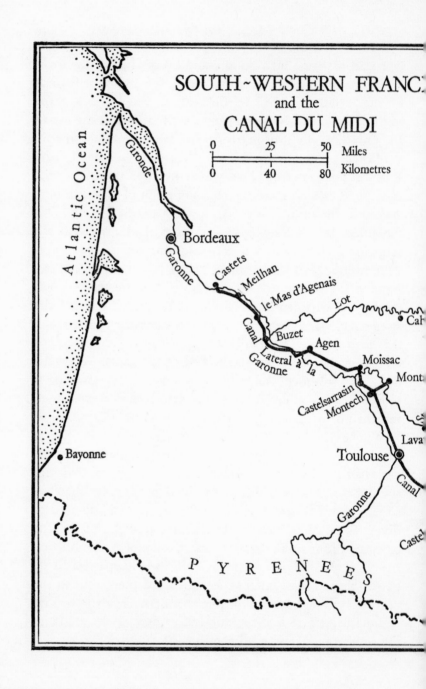

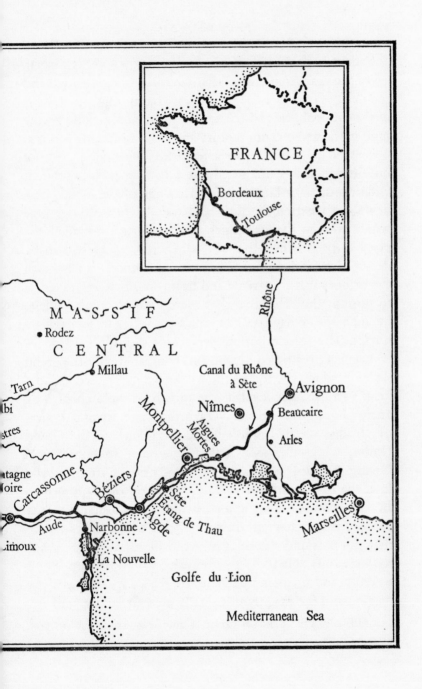

Riquet built a total of 139 road bridges over his canal. Over the years some have been rebuilt, while new road and rail bridges have been added in more recent times. Yet many of the original bridges still remain substantially as they were built in the seventeenth century and looking at them one is at once struck by two curious features. The first is that not one bridge that I can recall has a towing path beneath it wide enough for horses, a fact that must have made horse haulage an extremely slow and tedious proceeding. One wonders how the passenger 'packet boats' managed the point-to-point times they did when tow ropes had to be detached at each bridge. Where a path does exist beneath the bridges it is so narrow and headroom is so restricted as to prompt the belief that they were designed with haulage by men ('bow-hauliers') in mind rather than haulage by horses. The second curious feature of the original bridges is their very restricted dimensions compared with the grand scale on which the rest of the works were executed. With a height from water level to the crown of the arch of only 10 ft 9½ in,[1] they appear very little bigger than a typical overbridge on an English broad canal such as the Grand Union, while they are strikingly similar in appearance. Owing, perhaps, to settlement, some bridges are unusually low, the bridge known as de Capiscol on the approach to Béziers from Agde being a particularly bad example. Many modern barges have to dismantle their wheelhouses before they can pass under it unladen, while on my autumn voyage we were only able to do so after taking on board a posse of weighty and energetic Frenchmen who appeared as if by magic from the neighbouring railway marshalling yard.

1. Officially 3 m 25 and *not* 3 m 50 as quoted in certain guides to the French canals.

For the westbound traveller it is at Béziers that the Canal du Midi becomes exciting, both visually and from an engineering point of view, the most striking feature being the sensational 'new' crossing of the river Orb. How Riquet's original crossing of this river was contrived is shown by the contemporary map (plate 15). It will be seen that from the foot of the Fonserannes staircase the canal entered the river by a double lock at Notre Dame and left it again at a point 900 yards downstream, traffic being directed on this river section by timber booms secured to piles. Just below the downstream exit of the canal, a weir was built across the river and equipped with large sluices (*épanchoir*) the object of which was to enable the silt which built up in the navigable channel in time of flood to be scoured away. Despite this provision, the channel silted up repeatedly, while the height of the weir was such that it did not provide sufficient navigable depth of water in summer droughts. In the attempt to overcome this trouble, three new features were introduced. First, the crest of the downstream weir was raised so as to maintain a minimum depth of 6 ft in the river above in time of drought. After it had been swept away by a flood, this weir was rebuilt in the form of a series of stop planks between masonry columns. In the event of a heavy flood, some or all of these planks could be readily released, the planks being prevented from floating away by safety chains (see plate 16).

The old map shows a single pair of flood gates (*demi écluse*) on the canal at a point which was known as Moulin Neuf. To these a second pair was added at St Pierre by the mouth of the canal. These two pairs of gates enclosed 800 yards of canal which could be used as a refuge by boats in times of heavy flood. By means of sluices in the new St

Pierre gates, its waters could also be used to scour away silt deposited by floods in the canal entrance. Yet another attempt to cure the silting problem by more effectual scouring was the construction of a new weir equipped with powerful sluices just below the Notre Dame road bridge shown in the left hand corner of the map.

Looking at this map it is easy to see how craft bound for the Mediterranean could enter the river at Notre Dame and, guided by the booms, float down to the canal exit on the current. But how westbound traffic ever worked its way up the river is not known, unless walkways for bow-hauliers were provided on the booms. Evidently in later days a gap in the boom was formed beside the left bank, for a horse towing path was provided along that bank which enabled west-bound traffic to be hauled up the river from the mouth of the canal at St Pierre to Notre Dame bridge. There the horses were detached and crossed the river by the bridge, leaving the boats to float down on the current to the Notre Dame lock entrance. A similar towpath existed on the right bank between Notre Dame lock and the bridge, so that east-bound traffic could be hauled up to the bridge and then left to float down.

Over the years, however, the river defied and defeated every attempt to improve the crossing and in each year there were too many days when traffic was held up either by too much water or too little. In 1779, for example, owing to an exceptional flood, boats waited to cross for seventeen days on end. Though nothing was done, there was no lack of schemes for remedying this state of affairs. There were plans for high and low level aqueducts, the first coming from Vauban in 1739. He proposed an aqueduct 54 ft wide and 1295 yd long, taking off from the fifth chamber of the

L'Orbiel Aqueduct at Trèbes, designed by Vauban, *c.* 1686

The Fresquel Aqueduct, part of the Carcassonne diversion of 1810

36   Le Grand Bassin, Castelnaudary

37   Summer on the Canal du Midi

Fonserannes staircase counting from the top. This was judged too ambitious and too costly. There was even a serious proposal in 1756 to carry the canal *under* the river by a tunnel. The old double lock at Notre Dame was to be extended into a four-lock staircase leading directly into the mouth of the tunnel. To descend such a staircase into a dark cavern under a swollen river would certainly have been a highly dramatic canal experience, but how to ensure that the river itself never invaded the tunnel below?

It was not until 1854 that anything was done. Designs for the present aqueduct and its connecting canals were prepared by M. Magues, the Chief Engineer of the Canal Company, in April of that year, and by a decree of the Emperor Napoleon III dated 14 June the plans were formally approved and declared a work of public utility. Construction began immediately under the direction of M. Simonneau, Ingénieur des Ponts et Chaussées, and was completed in May 1856.

Approaching the tail of the lower of the two double locks that now lift the canal up to the level of the aqueduct, Riquet's old canal may be seen curving away to the left to join the river. The four chambers of the new locks are rectangular and the lowest of them has beside it a dry dock for the repair of barges. The lower end of the chamber is embanked and the height of the gates increased so that it can be over-filled for the purpose of floating barges on to or off the dock. This is a simple and ingenious arrangement such as I had never seen before. In between the two double locks, which are known as Béziers and Orb, is a spacious basin and quay, the Port Neuf, which replaced the old wharfage down by the river.

From the head of l'Orb lock the canal swings to the left

and immediately passes onto M. Magues's magnificent stone aqueduct of seven spans. Beneath the broad towpaths on either side of the canal are two walkways flanked by long arcades. Access to these lower walkways is via descending steps from the towpath, their entrances covered by iron trapdoors which, understandably but unfortunately, are kept locked. The walkways make a promenade fit for a Roman Emperor, but it is difficult to see what purpose they serve except possibly for occasional inspection and maintenance. One suspects that M. Magues may have made an architectural virtue out of the necessity to lighten the load on his pier foundations. There can be little doubt that Riquet would have approved the Orb Aqueduct as worthy of his great canal. Although built nearly two centuries later it has the same air of monumental permanence and grandeur that typifies the works of the reign of le Roi Soleil. At night when its arcades are now floodlit, the aqueduct presents a truly magical appearance. The spirit of the Renaissance was a long time a-dying.

Unlike the earlier aqueducts on the canal, the masonry trough of this Orb aqueduct was sealed against leakage with a layer of concrete. During the annual stoppage in 1951 part of this original sealing was cut out down to the masonry and replaced.

Crossing this aqueduct, the canal traveller is rewarded by a magnificent view of the ancient city of Béziers, its old grey houses climbing higgledy-piggledy up the steep slope of the hill from Notre Dame bridge to the castellated cathedral of St Nazaire and the fortified bishop's palace that crown its summit. Then this view of Riquet's birthplace is lost to sight as the canal swings round in a sweeping curve between an avenue of tall cypress trees to rejoin the old route near the foot of the celebrated lock staircase of Fonserannes. The point of junction between new and old is the seventh chamber of the staircase and for this purpose it has been enlarged until it resembles the round lock at Agde. Below it is the eighth chamber and the old arm of the canal, now disused, leading to Notre Dame lock and the river. This seventh chamber now has three pairs of gates so that, in theory, it could still be used as a lock, but in practice the third pair always stand open unless the new canal needs to be isolated for repair purposes. This means that today traffic only uses six steps of the Fonserannes staircase and not seven as stated in most guides to the French canals. Even so, it is still a very spectacular ascent and one that is now made remarkably easily and swiftly. For the operation of the gates and their sluices has now been electrified by the simple expedient of installing individual electric motors to drive the sluice gears and the pinions that actuate the gate racks. The lock keepers are responsible for these motorised operations, leaving crews free to concentrate on their boats. On reaching the summit and looking back over the Orb valley and the city of Béziers across the river, it is easy to understand why Riquet's critics thought him mad when he drove his canal to this point.

Only five pleasant, winding tree-shaded miles of canal separate the summit of Fonserannes from Riquet's second supreme folly, the famous *Percée de Malpas* beneath the Enserune ridge. Quite apart from the fact that it is the site of the first navigable canal tunnel in the world, this is an interesting and dramatic place for other reasons. On the top of the ridge of the hill above the canal are the traces of the ancient Iberian town of Enserune dating from the fifth century B.C. The line of the Roman road from Narbonne to

Béziers also crosses the ridge at this point. The top of the northern slope of the canal cutting at the approach to the tunnel directly overlooks the gleaming tracks of the canal's first great rival, the main line of the Chemin de Fer du Midi. The railway dives into a tunnel that passes beneath Riquet's canal tunnel on a long diagonal to emerge on its western side en route for Narbonne. Beyond the railway to the northward stretches the remarkable Étang de Montardy, once a shallow lake but now a great expanse of fertile land as flat as the proverbial pancake. It is remarkable because it was drained as long ago as the reign of St Louis in the thirteenth century. The drainage system, which was authorised by the Archbishop of Narbonne on 2 February 1247 and executed by monks, is still in use and is an astonishing feat of early engineering.

A sump was excavated in the centre of the *étang* and into this the whole area was drained by a system of radial ditches which extend like the spokes of a wheel from a central hub. The waters of the main drainage canal that was dug from this sump were carried beneath the Enserune ridge by an underground culvert 5 ft wide, 6 ft 6 in high and 1487 yards long at a level 51 ft below the base of Riquet's canal tunnel. This enabled the water to drain into the Poilhes and Capestang *étangs* which were not reclaimed until the end of the seventeenth century. Thus many generations of men from Iberians, Romans and medieval monks to Riquet and the railway builders have left their mark on this place.

So generous were the dimensions of Riquet's tunnel that there was originally a towpath on each side. That on the south side soon became ruinous and has long since disappeared. When it was being demolished a vertical shaft was discovered which proved to be either a working or a ventilation shaft of the old drainage culvert below. The resourceful canal engineers, ever anxious to take advantage of any additional means of ridding their long pound of surplus water, converted this shaft into an *épanchoir* through which flood water could fall dramatically and with hollow thunder into the culvert below. This shaft is shown in the section of Malpas tunnel reproduced in Général Andreossy's book (see plate 13). Today its mouth is covered by a stone slab which can be removed, but only if it becomes necessary to drain the floor of the canal tunnel completely.

It is remarkable that the western end of the tunnel has survived so long without lining or support, for the rock is so soft and friable that it crumbles to sand at a touch and the surface of the towpath hereabouts is covered with sand eroded from the tunnel walls.

From the cavernous mouth of Malpas the canal follows the southern slopes of the Enserune ridge before turning sharply northwards to round its western end at the village of Poilhes and Roquemalene hill, still hugging the contours. It was at the latter place that disaster struck in January 1744 in the shape of a major landslip. The hillside slipped down into the bed of the canal, blocking it completely for a distance of nearly 300 yards. The channel had to be excavated afresh and a massive retaining wall built to prevent a recurrence. Believe it or not, navigation was restored in fourteen days. The surprising thing is that so few incidents of this kind have occurred over the centuries, bearing in mind that the builders of the Canal du Midi were completely ignorant of geology and soil mechanics.

From this trouble spot, the long pound swings round to the west again to pass to the north of Capestang, high above the little town. From the old canal bridge beside the attractive canal 'port', a road leads steeply down through narrow streets towards the town's central square. This bridge, like that at de Capiscol below Béziers, is exceptionally low with the added disadvantage that, being situated on the long pound, there is no hope of lowering the water level temporarily.

It was at Capestang that the worst disaster in the long history of the Canal du Midi occurred. On 16 November 1766 a storm of a severity never known before burst over Languedoc and was followed by a period of severe frost and repeated falls of snow. Even in the middle of winter, ice and snow are rare in lower Languedoc. Swollen rivers caused immense damage and at Béziers the Orb swept away all the canal works at the river crossing. On the long pound the situation rapidly became critical. For despite the fact that water was pouring from every flood sluice and spill weir, such safety valves proved quite inadequate and the water level rose three feet above normal. Finally, the inevitable happened and the canal burst at Capestang. A section of the bank 46 yards long and 37 ft deep gave way and through this tremendous breach a raging torrent of water swept down through the town and into the Étang de Capestang.

Despite the appalling weather, the canal owners lost no time in coping with this disaster. They promptly raised a loan of 500,000 fr and a work-force of 10,000 men to repair the breach. It was judged impossible to restore the earth bank so it was decided to replace it by a massive masonry wall. Because no suitable material was available on the spot, every stone had to be extracted from remote quarries, hewn to shape and then transported to the site. Apparently enough water remained in the long pound to enable materials to be brought to Capestang in shallow draft boats, although each day the ice had to be broken to enable this traffic to continue. The unfinished wall was repeatedly covered with snow, and coal fires were kept burning night and day in order to dry out the mortar. By such desperate expedients the Capestang breach was successfully closed after three months' work.

To prevent a similar disaster happening again it was obvious that some additional means of relieving the canal of flood water must be provided. It was for this purpose that the son of a chief engineer of the canal, M. Garripuy, designed his ingenious *épanchoir à syphon*, or siphon sluice, which acted as an automatic water level regulator. This is built in masonry and consists of a rectangular culvert 18 inches high and about 3 ft wide which, in longitudinal section resembles a curved siphon pipe. The inlet of this culvert is two feet from the bottom of the canal, while at the apex of the curve of the siphon the height of the bottom of the culvert is equal to that of the normal water-level in the canal. This means that as soon as water-level exceeds the normal, water starts to flow through the culvert. In the ordinary course of events the device would continue to siphon until it sucked air through the inlet or, in other words, until there was only two feet of water left in the canal. To prevent this, M. Garripuy added an automatic valve (*venteuse*) which opens under suction and, by admitting air, stops the flow as soon as the canal level has returned to normal.

Two of these siphon sluices were built in the long pound at Capestang and Ventenac, the latter being double. So successful did they prove that a third was added at Fer-du-Mulet about two miles west of Capestang while a fourth was

built at Marseillette to regulate the level in the five mile
pound between that place and Trèbes.

The resourceful M. Garripuy also tackled another
trouble that has always bedevilled the long pound in summer – the prolific growth of water-weeds. He designed a
special weed-cutting boat, probably the first of its kind.
Beneath the stern of this boat three curved, horizontal
scythe blades projected radially from a central boss attached
to a vertical shaft having a pinion at its upper end. By means
of a toothed sector linked by two connecting rods to a
manually rotated crankshaft and flywheels, a rapid alterna-
ting motion was imparted to the scythe blades which
prevented them from becoming fouled with weed. Two
long screws enabled the frame on which the blades were
mounted to be adjusted for depth.

The course of the long pound west from Capestang
affords a classic example of a contour canal, its sinuous line
signalled from afar in a comparatively treeless landscape by
the continuous avenue of tall plane trees which march
beside it. So tortuous is it that the voyager practically
circumnavigates l'Ale Farm, two miles from Capestang, a
proceeding reminiscent of the similarly persistent view of
Wormleighton Hill farm from the summit level of the
Oxford Canal. In the blistering heat of summer, the plane
trees afford welcome shade and the seemingly endless vista
of their interlacing branches above the still water is en-
trancing. The only drawback is that the precautionary
flood-banks that Riquet's engineers raised, although no
longer continuous, cut off the view of the surrounding
country from those travelling in small pleasure craft. This
meant that I did not appreciate the beauties of the long pound
fully until I was able to view them from the deck of a larger

boat. This is some recompense for the anxiety caused by the low bridges.

West of Malpas tunnel, the line selected by Riquet for his high-level route means that for the westbound traveller the higher ground is usually on the right while on the opposite hand the land falls away to the Étang de Capestang, the Aude valley, and Narbonne. There is often a remarkable contrast between left and right. Where once was marshland there are now vineyards, orchards, groves of peach trees and fields of maize rolling away into the distance below the bank of the canal. Occasionally, if the atmosphere is sufficiently clear, the majestic snow-clad summits of the Pyrenees reveal themselves, an improbable backcloth, beyond the low hills bordering the Aude valley. With this scene the slopes that so often rise steeply from the right bank make a sharp contrast. For although they are often planted with vines, their soil is thin, stony and arid, appearing almost desert-like, particularly in winter or early spring when the rows of vines show only the black knuckles of their rootstocks. Here grow broom and myrtle, a few contorted olive trees or the black spires of cypresses. Occasionally one may see the solitary figure of a shepherd, leading his flock of brown, goat-like sheep. The villages that cling to these slopes: Argeliers, Ventenac, Paraza, Roubia and, particularly, Argens, each a close huddle of pantiled roofs and brown walls as bleached as old bones by centuries of strong sunlight, look as ancient as that Iberian settlement at Enserune. Certainly they can have changed scarcely at all since their inhabitants looked down in wonder on Riquet's men as they blasted their way through to Béziers. This landscape reminded me of the coloured illustrations of Palestine in a book of Bible stories that I used to look at as a child.

The Pont de Pigasse, near the farm of that name, bears above its keystone the sculptured armorial device of Languedoc, the Languedoc cross upon a shield surmounted by a crown. Another bridge carrying the D5 road over the canal near Argeliers is similarly embellished and doubtless in Riquet's day, proud Occitan that he was, many other bridges bore the same device. Near the Pigasse bridge a solitary lengthman's cottage stands beside a pair of stop grooves equipped with stop planks which can divide the long pound into approximately equal halves.

Following the second ornamented bridge near Argeliers there is the only considerable straight length on the long pound. Here, for once, there are no plane trees and instead both banks of the canal are thickly clothed with bushes interspersed with ancient, twisted parasol pines. After the civilised avenues of planes that have marched beside one for so long, the canal seems to have become very wild and remote. At the end of this straight the channel suddenly widens out and there on the left beneath the arch of the towpath bridge stretches the Canal de Jonction whose opening in 1776 at last supplied the missing link with the Aude and the Canal de la Robine and so satisfied the disgruntled citizens of Narbonne. A lonelier canal junction it would be difficult to find, the nearest habitation being the solitary lock cottage beside the top lock on the branch, 200 yards away. When we moored here one warm and still September night there was no sound at all other than the ceaseless song of the cicadas.

The Canal de Jonction from the Canal du Midi to the river Aude is only a little over 3 miles long (5 km) and there are seven locks falling to the Aude, the one at Sallèles being double, making eight chambers in all. On joining the river

it is necessary to turn directly upstream, then cross beneath
the cable that is suspended above the river and proceed
downstream, hugging the opposite bank, as far as the
entrance to the Canal de la Robine at Moussoulens lock,
just above the weir. This manoeuvre is necessary to avoid
the bar of silt that has built up in the centre of the river as a
result of the weir.

As originally built, the ancient Canal de la Robine was an
open cut, providing what was virtually a secondary channel
for the Aude between Moussoulens, Narbonne and the
Mediterranean at Port la Nouvelle. It followed a tortuous
course so as to reduce the gradient and therefore the strength
of the current. However, when the Canal de Jonction was
built its course was straightened and six single locks were
constructed. It is now fed from large sluices beside Moussoulens lock. Since Port la Nouvelle became an oil terminal
this route has been increasingly used by tanker barges
working between this port and Toulouse. It is also useful for
yachts travelling to or from harbours on the Mediterranean
coast of Spain, though owners of such craft should be
warned that draught is restricted to 4 ft 7 in (1·4 metres) and
that although, officially, headroom is the same as on the
Canal du Midi, in fact, clearance under the old bridges of
Narbonne is not more than 10 ft. Our attempt to enter the
Canal du Midi by this route in September 1971 was for this
reason defeated by the old Roman bridge at Narbonne, and
since the Robine is to all intents and purposes a river
navigation, there is no possibility of lowering the water
level. There was nothing for it but to beat an ignominious
retreat and continue along the coast as far as Agde.

A little to the west of the junction, the Canal du Midi
crosses the river Cesse by a masonry aqueduct of three

arches, a centre span of 60 ft and two side spans of 48 ft
each. It was designed in 1686 by Sébastien Vauban and built
by Antoine Niquet. This aqueduct replaced a curved dam,
672 ft long and 30 ft high, which Riquet had built across the
river in order to create a level crossing. When this dam was
demolished and the aqueduct built in its stead, Vauban and
Niquet had to make good the water supply which the long
pound had hitherto received directly from the Cesse. For
this purpose the Roupille diversion dam was built across the
river at the village of Mirepeisset and from above this the
water is fed to the canal through the 3000 yard long Mirepeisset feeder. Terraced along a steep bank of the river, this
enters the canal immediately to the west of the aqueduct
where it is controlled by a sluice. Although it was estimated
that this feeder delivered 9,773,060 gallons (44,423 cu m)
of water to the long pound every twenty-four hours it
was insufficient to make good that consumed by traffic using
the new branch canal to Narbonne. Hence the construction
of the Lampy reservoir.

From the Cesse aqueduct it is little more than a mile to le
Somail, a true canal village which was the canal port for
Narbonne before the branch canal was opened. Here the
packet boats once lay for the night and carts brought the
produce of the district for shipment. Although it boasted no
large hotel for the reception of passengers, I was reminded
of that other one-time packet port at Robertstown on the
Grand Canal of Ireland. Beside the high-arching bridge
which carries the little village street over the water is the
quay and warehouse of the old port, disused now and, like
some abandoned country railway station, watching the
traffic pass it by. On the right just beyond the bridge there is
an attractive row of houses flanked by cypresses and

separated from the canal bank by a strip of greensward on which grow the ubiquitous plane trees. We tied up for the night to these plane trees on our eastbound voyage in the spring and thought there could be few more pleasant mooring places on the long pound.

The direction of the canal is here south-westerly and as a result the canal and the Aude converge until the former reaches a position on the north slope of the river valley which it is to follow all the way to Carcassonne. On the left is the flood plain of the Aude, which explains why the villages of Ventenac, Paraza, Roubia and Argens are all sited on the high ground to the right of the canal. Between Ventenac and Paraza the canal makes an excursion reminiscent of the old line of the Oxford Canal between Hawkesbury and Braunston, turning up the flank of the valley of the tributary Répudre until an easy crossing place can be found and then returning down the other side of the stream. At the apex of this diversion we stopped and clambered down the steep bank to photograph and to pay homage to Riquet's 'premier' Pont-Canal of 1676 with its single modest arch of 35 ft span.

Argens lock, between Roubia and Argens, marks the end of the long pound and hereafter the locks are seldom more than a mile apart all the way to Marseillette. Beyond Argens the canal approaches the river closely in order to pass through that narrow defile beneath the rock of Pechlaurier that provoked so much controversy between Riquet's allies and enemies. Looking up at the steep and crumbling slopes that rise above the canal on the right, one marvels that this has never been the scene of a landslip such as occurred at Roquemalane. Once clear of this high ground the canal, following the course of the river, bends sharply to the west to reach the tail of the double lock at L'Ognon. It was immediately below this lock that the Ognon stream was formerly crossed on the level. There is now a small aqueduct (built in 1826–7), but in case a high flood should overtop it and flow into the canal, the floodgate that originally protected the crossing on the Argens side has been retained. Known as the Garde d'Ognon, it is the only example left on the canal.

Similarly, beyond the little town of Homps with its spacious quay there is, near a place called Métaierie du Bois, on the pound between Jouarres Lock and Laredorte, a surviving example of the type of long flood relief weir with towpath carried over it on a series of arches such as were built by Riquet at his river crossings. This weir has been retained in case the Argent-Double stream, which is culverted under the canal at Laredorte, should flood and flow into the canal.

Between Puicheric and Marseillette the canal begins to climb more steeply by a series of double locks and one triple staircase until the Marseillette pound is reached after passing through the single lock of that name. From this lock there is an unexpected and splendid view to the north across the broad level of the Étang de Marseillette to the heights of the Massif Central which, when we passed in the early spring, were still capped with snow. Until the beginning of the nineteenth century the Étang de Marseillette was still a vast lake of shallow water of an average depth of 9 ft. The first proposal to drain it was put forward in 1747, but nothing was done until the Revolution when it was confiscated from its owner, the Comte de Caraman. It was then sold upon condition that the purchasers would drain it within four years. Work was started in 1804, when a main drainage channel was dug to join the Aude at Puicheric. In the process, a new aqueduct had to be built to convey this 'rigole d'étang', as it is called, beneath the Canal du Midi just below the tail of l'Aiguille double lock.

Beside the canal in Marseillette village there is a good example of the communal washing places provided at many points along the canal for the benefit of village housewives. Sometimes these consist merely of a slope of concrete on which the women kneel to pummel their washing, but here, as elsewhere, they are not only provided with a roof over their heads but with a concrete trough, its sides just above water level, in which they can stand instead of kneeling. They are apt to eye passing boats with some apprehension, however, for any excessive wash may overtop their standing trough and it is therefore obligatory to slow down when passing these washplaces.

The five-mile pound between Marseillette and Trèbes is one of the most beautiful on the canal. From the shade of the varied trees in the grounds of a château on the right, one looks across the valley of the Aude towards the dark l'Alaric mountains beyond and, if one is lucky, to the still higher skyline of the Pyrenees. It is here, between the two bridges of Millepetit and Saint-Julia that the half-distance mark between the Étang de Thau and Toulouse is passed. The end of this pound at the foot of the triple staircase lock at Trèbes also marks the end of Riquet's second contract and the beginning of the first.

SEVEN

The Canal Today –
Trèbes to Toulouse and
the Lateral Canal

At Trèbes there is a meeting of the ways. Here the canal and a secondary main road (the N 610) coming direct from Béziers begin to converge with the main line of the Chemin de Fer du Midi and the great trunk road across southern France, both approaching from Narbonne on the south bank of the river Aude. A glance at a map makes it easy to understand why there was so much difference of opinion over the route for the second section of the canal and why Riquet's final decision to bypass Narbonne by keeping to the high ground north of the Aude and driving directly towards Béziers should have encountered so much bitter opposition.

Once Carcassonne is passed, however, canal, railway and trunk road parallel each other closely up the valley of the Fresquel towards the Col de maurouze and thence down the Lers Valley to Toulouse, thus emphasising the fact that on this first section of the canal, topography determined the line of the canal beyond serious dispute.

From the head of the triple lock staircase at Trèbes, the canal swings sharply through the heart of the little town, under a bridge carrying the busy road to Béziers and past the disused town quay to come, through an avenue of

38   The *Ste. Germaine*, a converted horse boat with its original wooden tiller

39   A winter scene at Toulouse

40 Les Ponts Jumeaux, Toulouse, showing, *from left to right*, the Lateral, Midi and Brienne canals, also the upper part of the Port de l'Embouchure. The marble bas relief by Lucas can be seen between the twin bridges

41 Barges awaiting the end of a 'stoppage' in the Port de l'Embouchure, Toulouse

cypresses, to the aqueduct over the river Orbiel. As in the case of the Cesse, Riquet's crossing of the Orbiel on the level caused chronic silting up of the canal each time the river flooded and in 1686–87 Antoine Niquet demolished Riquet's dam and replaced it by the present aqueduct which was designed by Vauban. This has three equal spans of 36 ft and is set between very broad abutments which, like the face of the aqueduct itself, are empanelled and carry a single band of bold stone moulding. This simple but effective treatment makes it a more satisfying architectural composition than the Cesse aqueduct. Both these aqueducts of Vauban's, like Riquet's prototype the Répudre, are so massively built that, unlike most English aqueducts of similar construction, their side walls require no cross-bracing to resist outward water pressure. When we stopped to photograph the Orbiel aqueduct it was a warm day in early spring and little lizards were already sunning themselves on its weathered stone. The small river flowed with scarcely a sound beneath the arches, but the shoals of gravel and the sizeable tree trunks which littered its bed reminded us that it was not always so docile and placid. So that the Orbiel water would not be lost to the canal, when the aqueduct was built a feeder channel half a mile long was constructed from the river. This enters the canal at a point about 300 yards to the west.

Between Trèbes and Carcassonne the Aude makes a great loop to the northward with the effect that the canal, to keep its position on the north side of the valley, is forced to make an even greater loop, rising the while through the two attractively situated locks at Villedubert and l'Evêque. Then the canal straightens out and, at the end of an avenue of trees, the traveller sees the two locks, a single quickly followed by a double, that lift the canal over the river Fresquel. The present Fresquel aqueduct forms part of the first major deviation of the canal away from Riquet's original line in order to serve the city of Carcassonne. After years of lobbying by the city, the parliament at Toulouse finally agreed to undertake the work in 1786. The Revolution then caused the project to be postponed before any work had been done. The regional parliament was dissolved, but the new Revolutionary Central Government finally authorised work to proceed, appropriating 200,000 francs per annum from canal income in order to finance it. The first stone of the new aqueduct was laid in 1802 and the whole of the deviation was completed and opened in 1810. Thousands of Austrian and Prussian prisoners of war were employed in digging the cuttings and the total cost was estimated at 2,220,000 francs. The names of the engineers responsible for the design and execution of the works are not known.

Although the total length of new canal only amounted to about $3\frac{1}{2}$ miles, it involved very heavy engineering works, completely changing the topography in the vicinity of Riquet's original Fresquel crossing which has disappeared as though it had never been. This is because the Fresquel was diverted from its old bed into an artificial channel dug through rising ground to the west of it. It is this fresh channel that the new aqueduct crosses, and when the river was diverted into it, part of its old course was converted into a canal feeder. Between the outfall of this feeder and the river Aude the only thing which marks the old course of the Fresquel is the single stone arch of the old Pont Rouge bridge which once spanned it but now stands high-and-dry amid the fields to the left of the canal to mystify the passer by. All trace of Riquet's weir and eighteen-arch towpath bridge has vanished.

The original line of Riquet's canal followed the Fresquel valley throughout. It diverged at Pouillariès, a mile to the west of Carcassonne, passed through a triple staircase lock at Foucaud followed by a double lock at Villaudy, and rejoined the present route above the single Fresquel lock. In other words, this single lock is a reconstructed original whereas the double lock immediately above it is, like the aqueduct at its head, a part of the Carcassonne diversion works.

Until the completion of the Orb aqueduct at Béziers, the aqueduct over the Fresquel was the largest single engineering work on the canal, carrying both canal and road over the new channel of the river. The 38 ft span of the three equal arches is little more than that of the Orbiel and substantially less than the centre arch of the Cesse, but the fact that the height of the crowns of the arches above river level is so much greater makes this Fresquel aqueduct a far more imposing structure. With its deep cornice directly below the parapet, its design is reminiscent of John Rennie's aqueduct over the Lune at Lancaster which was completed a few years earlier (1797).

Immediately beyond the aqueduct the canal swings sharply to the south and heads directly for Carcassonne, skirting the high ground on the right which divides the valleys of the Fresquel and Aude. Beyond the single lock at St Jean the land on the left falls away to afford the canal traveller a splendid view of *la Cité*, the ancient walled city of Carcassonne dominating the high ground on the south bank of the Aude. Or rather it would be a splendid view were it not marred by the cat's-cradle of the typical

French urban wire-scape in the foreground. When first seen thus in the middle distance the curtain walls, the turrets and the towers of *la Cité* look as improbable as a film set on a studio lot. In fact, such an impression is not too wide of the mark for *la Cité* has often been used as a film background, while its present appearance owes a great deal to its careful nineteenth century restoration by Viollet-le-Duc. All summer long tourists swarm about its towers like so many wasps round jam pots, while the ancient city that its walls protected is now dedicated solely to the tourist trade. Fortunately, there are in France many less overtly spectacular but equally ancient places that are seldom visited by sightseers and where life goes on much as it has always done, quite unaffected by the self-conscious blight of tourism. They owe this happy immunity at least in part to the existence of 'five star attractions' like *la Cité* of Carcassonne and for this we should be duly grateful to the painstaking M. Viollet-le-Duc.

The bridge across the tail of Carcassonne lock, carrying the main street linking the 'new' city with its railway station, shares with the de Capiscol and Capestang bridges the dubious distinction of being exceptionally low. Because it dates from the beginning of the nineteenth century and there has obviously not been any subsequent settlement, this failure to provide adequate headroom seems inexcusable. However, in this case the pound between St Jean and Carcassonne locks is short and therefore can easily and quickly be lowered to provide those vital extra inches of clearance.

Immediately above the lock is the capacious basin and quay that is the 'port' of Carcassonne. Because it is within walking distance of both 'new' and ancient cities and is seldom or never used by trading barges, it now makes a convenient mooring place for visiting pleasure craft. West of this basin, the canal immediately enters the long and deep cutting by which it passes through the high ground in order to regain the Fresquel valley. It is certainly a formidable excavation for its date and it is not surprising that this short diversion should have entailed such a prodigal expenditure of time, labour and money. It inevitably prompts the speculation as to how Riquet would have tackled the job if the citizens of Carcassonne had come up with the money he needed in order to bring his canal into the city. Would he have dug such a deep cutting? Or would he, as Brindley did in similar circumstances, have resorted to tunnelling? There would appear to be no easier choice of route.

A single cottage beside the towpath marks the end of the Carcassonne diversion at Pouillariès, but the railway, which passes on an embankment close beside the canal at this point, has here obliterated all traces of Riquet's old canal. From the other side of the railway, however, it is visible for about half a mile, the site of the old triple staircase lock at Foucaud being just to the east of the trunk road crossing. The flood relief channel from the *Épanchoir de Foucaud* follows the line of the old canal for some distance until it turns aside to empty into the Fresquel.

The pounds between Lalande, Villèsèque and Béteille locks are both as tortuous as the long pound, in fact some of the bends are even more acute. On one of them near the hamlet of St Michel the steep bank on the inside of the bend has been so scoured away by the wash of passing traffic that a wide shoal has formed, making a trap for the unwary navigator. The steerer of a boat that is deep in draught who does not keep well to the outside is apt to find himself heading straight for the bank because the stern of his boat is embedded in soft mud and refuses to answer the rudder. This sort of disconcerting incident can occur all too often on English canals but is extremely rare on the Canal du Midi thanks to a systematic programme of dredging. Consequently this hazard may have disappeared before these words appear in print.

Beyond Béteille lock the canal passes under a bridge known dramatically as the Pont du Diable. The road this bridge carries, now the D33, is a Roman highway, the ancient predecessor of the present east to west trunk road that runs along the north side of the valley. Soon the main railway line also crosses the canal, heading, like the Roman road, for the ancient town of Bram. Riquet's canal by-passes Bram, but the little canal port of Bram was built at a point where a road crosses the canal about a mile from the town. It has long been disused and stands lonely and deserted now, just a small grass-grown quay, and a shuttered grey stone building beside the grey arc of the bridge. When we moored there briefly for a meal I was again reminded strongly of Ireland. It seemed to me to resemble many of those forlorn little canal harbours which I visited long ago on Ireland's Grand and Royal canals.

So straight has the course of the canal now become, so flat the lands on either hand and so inconspicuous the little river compared with the powerful Aude that one is no longer at all conscious of the fact that the canal is still following a river valley. But an insignificant cast-iron plaque about three feet above canal water-level on the arch of Bram bridge reveals what even a minor French river can do. It records the height of a Fresquel flood in January 1916. On that occasion

there surely must have been fears for the safety of the dams
and feeder system up in the Montagne Noire because all the
streams that supply Riquet's *rigole de la montagne* are tributaries of the Fresquel.

The nearer the canal approaches the headwaters of the
Fresquel the more steeply it has to climb. In the 10 miles
between the Port of Bram and Castelnaudary there are
nine single locks followed by the Vivier triple and the Gay
double locks until finally the quadruple lock staircase of
St Roch lifts the traveller into the great basin at Castelnaudary. As at Fonserannes, the sluices and gates of the
St Roch staircase are now electrically operated but, unlike
the former, these are operated from a switch panel in a high,
glass-sided control cabin that commands a view over the
whole staircase. As a result, boats work their way up or
down surprisingly quickly as though assisted by a super-
efficient team of phantom lock keepers. Gates open or close
and sluices lift and shut as though by magic. To a canal
boater of the old school who prefers to have everything
under his own control, even at the expense of callused
hands, it is all rather disquieting.

Le Grand Bassin is truly grand and to sail out on to this
broad expanse of water from the top chamber of the St Roch
staircase is one of the highlights of a westbound voyage
through the canal. On the south side of the basin is an old
storehouse, said to date from the days of Riquet, and
beside this the British firm of Blue Line Cruisers has
established its base. Directly opposite, the old town of
Castelnaudary stands mirrored in the still water, its houses
rising one above the other almost sheer from the water's
edge. For this is not the canal port; that is situated beyond
the steeply arched bridge at the western end of the basin.

As explained previously, the purpose of this basin is to act as a reservoir for the St Roch locks. Passing traffic wishing to tie up for the night usually lies along the quay wall of the old port from which narrow streets lead directly up into the town.

Leaving Castelnaudary, the canal continues its climb towards the summit, first by two single locks and then more steeply by the Laurens triple staircase and the Roc double lock. From the head of Roc lock it is not more than a kilometre to the single lock which was originally called Médecin but is now known as *l'Écluse de la Méditerranée* because it marks the eastern end of the summit level. The summit pound itself is only three miles long and situated exactly half way along it is the little village of le Ségala. Like le Somail, this is a canal village which reminded me of Robertstown although in this case there appears to be no historical reason for its existence. Le Ségala was never a 'packet port', but its situation suggests that in the early days of horse haulage it may have been a more convenient overnight stopping place than Castelnaudary for traffic other than the packet boats.

As they approach the Col de Naurouze, railway, canal and trunk road converge, the railway on the left hand and the road on the right. Were it not for a glimpse of the Riquet obelisk perched on the Stones of Naurouze away to the right, the canal traveller could easily pass this historic place unawares. For the canal runs between the green slopes of a shallow, tree-shaded cutting where that vital artery of the canal, the *rigole de la plaine*, makes its inconspicuous entry. To see the site of Riquet's great octagonal basin or the remains of the old locks beside the mill and lockhouse it is necessary to stop and go ashore and those who do so will be richly rewarded. One might expect the water in this summit pound, fresh from the Montagne Noire, to be crystal clear, but, though it may be pure enough, it is usually cloudy with finely suspended silt.

Round a gentle bend just beyond the site of the Naurouze basin the top gates of the western summit lock come into sight. This lock was originally named Montferrand after a nearby village, but is now known as *l'Écluse de l'Ocean* because it marks the beginning of the long descent towards the Atlantic. Altogether there are eight single and eight double locks between the summit and the Port de l'Embouchure at Toulouse.

About a mile to the west of Renneville single lock, and between Villefranche and Gardouch, the canal makes a sweeping S bend, crossing over the headwaters of the little river Lers, to gain the south side of the Lers valley which it then follows more or less directly all the way to Toulouse. It is interesting to notice the old mill buildings beside the double lock at Castanet, for this was one of the four locks between the summit and Toulouse where a side weir was originally provided by Riquet for the purpose of driving a mill. The sites of the other three locks are all within the city of Toulouse and no trace of their mills has survived.

Once over the summit level, the canal traveller becomes very conscious of the fact that he has passed from a Mediterranean to a more temperate Atlantic climate. This landscape of green pastures interspersed with arable fields is much more reminiscent of England, particularly when, on approaching Toulouse, brick canal bridges take the place of stone. By contrast, the semi-tropical landscape of the long pound with its vineyards, its olive and cypress trees and its ancient hilltop villages already seems so remote in recollection that the canal traveller feels like a voyager returning from foreign parts. Only the fact that it is maize and not wheat or barley that grows in the arable fields reminds the northerner that he is a long way from home.

The Canal du Midi loops its way round the eastern boundary of the ancient city of Toulouse which lies between it and the Garonne with the effect that little of the city that can be seen from the water dates from before the nineteenth century. Nevertheless, to travel through any city, ancient or modern, by canal is always a fascinating experience and Toulouse is no exception. It is surprising how effective even a few feet of still water can be as an insulator against the high-tension currents of modern life. On the perimeter of the city where until lately there were green fields, the tree-lined canal passes through the campus of the University of Toulouse, its tall modern buildings rising up on either hand. Because the canal is lock-free between the University and the city centre, some enterprising character was operating a shuttle service by water-bus.

The canal banks soon become more densely built-up, while ahead the vista of canal bridges resembles a continuous arcade. On the left the entrance to the canal basin known as the Port St Sauveur opens up. For those wishing to visit the old part of the city or Riquet's resting place, the cathedral of St Etienne, St Sauveur offers a more convenient mooring than the terminal Port de l'Embouchure. After passing beneath six more bridges the canal traveller suddenly finds himself in the middle of a broad and busy tree-lined boulevard, the imposing station building of the Chemin de Fer du Midi on the right, shops and cafés on the left and a swirling torrent of city traffic on either hand.

Ahead are the gates of Bayard double lock, the first of the four, two single and two double, by which the canal descends to its terminal basin. It was here that we moored for the night on our arrival in Toulouse, Bayard lock having just closed. After days and nights spent in the quiet of this essentially rural canal, it was a dramatic contrast to tie up for the night in the heart of the fourth largest city in France, searching for mooring bollards amongst the parked cars that lined the cobbled quay.

On descending the locks it is surprising to find Riquet's canal sandwiched between the two carriageways of a modern underpass, both descending to a level lower than that of the canal in order to gain the necessary headroom. There is talk of diverting the Canal du Midi from its old course through the city to make way for a new road. After a night spent in its midst, we could appreciate that Toulouse has an acute traffic problem, but it is to be hoped that some less drastic solution may be found. For in any modern city a quiet street of water with its slow-moving traffic is a priceless asset, not lightly to be sacrificed to the motor car.

On leaving the last lock at Bearnais, the bridge leading into the Port de l'Embouchure appears at the end of a final straight, tree-lined avenue of water. With the completion in 1776 of the short Canal St Pierre, or Brienne Canal, which enters the basin close beside it, this last bridge over the Canal du Midi has become one of a pair known as *les Ponts Jumeaux* (the twin bridges). The shared abutment between them bears an enormous bas relief in marble carved in 1775 by the sculptor François Lucas in which allegorical figures celebrate Riquet's great achievement. A central figure representing the Province is seated on the prow of a vessel bearing the arms of Languedoc. A fierce looking bearded fellow to the left of the ship, looking rather like a Father Neptune who has mislaid his trident, appears to be making a pass at the nubile and naked young woman on the right. It is said that these two figures represent the canal and the river respectively, but the contrast between the peaceful canal and the untamed savagery of the Garonne makes such an attribution singularly inappropriate. It seems much more likely that the sculptor intended to depict the union of the Atlantic with the Mediterranean.

The history of the Brienne Canal is as follows. A mile up-river from the old entrance lock to the Canal du Midi, the Garonne was closed to further navigation in the thirteenth century by the construction of the Bazacle dam; 900 ft long and crossing the river on a long diagonal, its purpose was to provide water power for the first of the celebrated mills of Bazacle. After withstanding the violence of the Garonne for centuries, this medieval dam was completely carried away by an exceptional flood in the year 1709. Under the direction of an engineer named Abeille, a new dam, 16 ft high and nearly 70 ft thick, was built across the river. It stands to this day having, in 1722 and 1835, withstood two of the worst floods ever known on the Garonne. In 1768, construction of the Brienne Canal was commenced by the Province of Languedoc under the aegis of Cardinal Loménie de Brienne with the object of linking the Port de l'Embouchure with the upper Garonne and so providing small river craft with a means of by-passing the Bazacle dam. By a decree dated 10 March 1810 its ownership was transferred to the Compagnie du Canal du Midi. Although its traffic dwindled and finally ceased, it provided the engineers of the Lateral Canal with a ready-made feeder, using the Bazacle Dam as a diversion weir, a function it still performs very effectively although there is now talk of using its course for a new road, presumably by converting it into an underground culvert.

With the completion of the Garonne Lateral Canal, a third bridge was added to *Les Ponts Jumeaux*. This spans the entrance to the Lateral Canal, is set at right-angles to the Midi Canal bridge and in close proximity to it. This arrangement may have been all very well in the days of horse haulage, but with the introduction of power-propelled craft it became fraught with hazard as will be obvious from a glance at plate 40. Unless barges debouching into the basin from either of the canals proceed with great circumspection there is liable to be a thunderous collision accompanied by much profanity. So tight is the turn from one canal into the other that traffic usually proceeds into the basin and turns about there before heading off in one direction or the other.

The Lateral Canal is virtually a latterday extension of the Canal du Midi. It is said that the idea of a canal which would by-pass the turbulent Garonne was first advanced by Pierre Paul Riquet himself. It was subsequently advocated by other engineers including Vauban and, in the eighteenth century, by Chief Engineer Garripuy, Director of Public Works in Languedoc and the designer of the Lampy Dam. But nothing was done and the man finally responsible for getting the project off the ground was de Baudré, Divisional Inspector of Highways and Bridges. The names of the engineers responsible for the design and execution of the canal do not appear to have been recorded, but probably de Baudré was responsible for the original survey and contract drawings since it was thanks to his initiative that a private company submitted a successful tender for its construction about 1830. After this company had twice gone bankrupt, however, the government purchased the unfinished works in 1838 and carried them through to completion under the supervision of their own engineers.

The canal was opened from Toulouse to Agen in 1850, while the second section from Agen to its junction with the tidal Garonne at Castets-en-Dorthe was completed in 1856. It is 120 miles 5 furlongs (193 km) in length with fifty-three falling locks including the two tidal locks at Castets. There are no staircases or double locks on the canal. Lock chambers are of the normal rectangular pattern and, as originally built, measured 100 ft between gates and 20 ft wide. The lock gates, with no balance beams and rack operation are of the same type as those now used on the Canal du Midi. It is possible that this design was first evolved for the Lateral Canal and, proving successful, was subsequently introduced on to the Canal du Midi as existing timber gates of the old design required replacing. The lock cottages bear the same attractive cast-iron name and distance plaques as those on the Midi except that here, owing to an unfortunate misunderstanding between the canal engineers and the foundry responsible for them as to which side of the canal the cottages were to be built, some of the distance figures have been transposed.

The Brienne Canal feeds the Lateral Canal with water as far as Agen and, except in the driest season, the supply is such that the side-weirs beside the locks run with considerable force, many of them driving small hydro-electric generating sets. Because the inlets and outlets of these side-weirs are badly laid out, being at right-angles to the canal, even quite heavy craft are liable to be deflected from their course when slowing down to enter the locks, either by the pull from the head of the weir or by the cross-current caused by its outlet at the lock tail. This makes it difficult to enter locks without bumping the gates or side walls and pleasure craft using the canal should be well equipped with fenders.

The channel of the canal is somewhat wider and deeper than that of the Canal du Midi being 60 ft (18 m) wide at the surface with a bottom width of 38 ft 4 inches (11·5 m) which allows a draught of 6 ft (1·8 m). The official figure for head-room is 11 ft (3·30 m), but most over bridges allow considerably more than this being, not arched bridges, but single bow-and-string spans of reinforced concrete construction which obviously post-date the canal.

The first reaction of a traveller leaving the Midi for the Lateral Canal could scarcely fail to be one of disillusionment accompanied by a strong desire to put about and return by the way he came. For a duller stretch of water than the first 35 miles from Toulouse as far as Castelsarrasin it would be difficult to conceive. In contrast to the vagrant windings of the Canal du Midi, it cuts as straight as a Roman road through a flat and featureless countryside, closely accompanied for most of the distance by the trunk road and railway line. Though the river Garonne is never far away to the south, it is not visible and the only feature worth noting is the junction of the Montauban Branch at Montech. Seven miles long with ten locks falling from the main line, this branch was opened in 1843 and originally made a junction with the River Tarn at Montauban. Soon after its opening a packet boat service began operating between Montauban and Toulouse. Despite the number of locks on the route, this apparently competed successfully with the road coaches. The junction with the Tarn is now stopped off, but those

inclined to visit the old town with its splendid fourteenth-
century bridge over the river can still turn aside down the
branch and moor in a deserted terminal basin.

It is after passing Castelsarrasin that the Lateral Canal
suddenly and dramatically changes its character and after
this its interest seldom flags. Emerging from a shallow
cutting, the traveller suddenly finds himself sailing high
above the river Tarn on a magnificent aqueduct of thirteen
spans. On the left is a weir and beside it is one of the derelict
locks of the old Tarn Navigation. On the right, the main
railway line between Toulouse and Bordeaux crosses the
river by a long bridge which replaces one that was completely
swept away by a great flood in 1930. It was a tribute to the
canal engineers that their aqueduct should have stood firm
even when the wreckage of the railway bridge was swept
down upon it. It was even more remarkable that the engineers should have decided to use the aqueduct for a temporary rail crossing while the railway bridge was rebuilt.
The channel over the aqueduct was sufficiently broad to
enable the towpath to be temporarily widened to admit a
single line of rails, the canal being closed for six weeks while
this work was done. Two single-line spurs, each a mile
long, linked this improvised crossing to the sundered main
line. Thereafter the aqueduct carried both rail and canal
traffic for two years until the present railway bridge was
opened. Although movement of trains across it was restricted to a walking pace, it is nevertheless remarkable that the
aqueduct should have withstood this considerable additional
off-centre rolling load for such a long period without
sustaining the slightest damage.

The course of the canal until now has been almost due
north, but once the Tarn is crossed it turns sharply westward,

42, 43 Ancient and Modern: *top*, a staircase lock on the
Canal du Midi; *below*, a lengthened and power-
operated lock on the Lateral Canal at Agen

44   Day's work done: Midi Canal barges at moorings

descending into the river valley by three locks to traverse the delightful little riverside town of Moissac. From Moissac lock the prospect ahead is singularly beautiful, the canal passing between smoothly sloping grass banks flanked by the old buildings of the town, which now look down upon a street of still water. Immediately below the lock, a disused flight of locks leads off to the left to join the river. One feels that the construction of the canal must have caused as much consternation to the inhabitants of Moissac as would the coming of a new motorway today, for it has literally split the town in two. For some hundreds of yards the canal runs between retaining walls down the centre of what must once have been the town's wide main street, the two narrow thoroughfares that were left on either side being linked by swing bridges operated by bridge-keepers. But at least the canal engineers did not, as the railway engineers did, threaten the existence of Moissac's ancient abbey church of St Peter with its cloisters. Very fortunately this threat was averted and one of the finest examples of Romanesque architecture in France was reprieved. The reason why Moissac thus became the target for two rival transport routes is that the town completely occupies a narrow defile between the Tarn and the steep slopes of the hills that here enclose its valley to the north.

Just beyond Moissac is the confluence of the Tarn with the Garonne and for about a mile thereafter the range of hills to the north forces canal, railway and road into close company beside the great river. Then at the village of Malause the hills fall away to the north and the river inclines to the south, leaving the three transport routes free to fan out on a north-westerly course through flatter but pleasant country in the direction of Agen.

From Moissac to Agen is 26 miles. The canal winds between endless fields of maize and at one point a completely new canal port and a depot with huge silos has been created in the midst of these fields for the shipment of the grain by water. Agen is the largest centre of population on the Lateral Canal west of Toulouse, but it is a not particularly interesting or attractive city at the best of times and the canal, looping round to the north of it like a defensive moat, merely skirts its backyards. The quickest way to reach the centre of the city from the canal is to cross a long steel footbridge over an extensive railway marshalling yard. It is all rather depressing, but the Lateral Canal seems to have a way of reviving the spirits of the traveller just as they are beginning to flag. Having worked its way round to the west of the city, the canal suddenly strides across the wide Garonne on a second splendid masonry aqueduct. This is even larger than that over the Tarn, being over a quarter of a mile long (539 m). These Tarn and Garonne aqueducts are without any doubt the greatest works of civil engineering on the water route between the Mediterranean and the Atlantic. The bases of their troughs are formed of cast-iron plates keyed into the masonry of the side-walls and abutments. This was the method introduced by Thomas Telford when he designed his aqueduct over the Ceiriog at Chirk on the Ellesmere Canal, which was completed in 1801. In the use of cast iron for canal construction, British engineers undoubtedly led the world and it is interesting to speculate whether there was any borrowing of ideas on the part of the engineers of the Lateral Canal. If there was, it was a fair exchange for all that Riquet and his engineers had earlier taught the world by their example.

Although the main road and railway line continue along

the north bank of the Garonne west of Agen, a glance at a physical map of the area is sufficient to reveal why the canal engineers chose to cross the river at this point. For whereas to the north hills rise steeply from the river as at Moissac, on the south side the valley floor is wider for many miles downstream. The canal descends to this floor by a flight of four locks, the first of which, Agen Lock, is situated immediately beyond the south end of the aqueduct. This is the last lock flight on the canal. For the remaining distance to Castets the locks come singly and widely spaced, apart from a pair just beyond the small aqueduct over the tributary river Baise. Just beyond the bridge at the foot of this flight of four locks a channel from the river fed by the Beauregard diversion weir, flows in on the left. As far as this point, a distance of 69 miles from Toulouse, the canal has been wholly dependent for its water supply on the Canal de Brienne. This Agen feeder, like the Brienne, was once navigable, affording communication with the river through a flood lock.

Just before the little village of Buzet-sur-Baise there was yet another communication between the canal and the Garonne. Here two disused locks lead down into the river Baise which was formerly navigable from the Garonne up to the little towns Vianne, Lavardac and Barbaste to the south of the canal. Hereabouts, as on the first stretch after Toulouse, the pounds are long and dead straight but with the difference that the landscape is now deeply rural, green, well-timbered and dotted with many small villages so that the journey never becomes monotonous. Gradually, however, the high ground that has been marching along to the south begins to converge, forcing the canal to keep closer company with the river until, between the hamlet of

Lagruère and the village of le Mas-d'Agenais, it is terraced along a steep bank directly above the Garonne. Then the river loops away northward to appear once more at Caumont and then again, most spectacularly, at Meilhan-sur-Garonne. Here the village, with its old houses and steep winding streets clings precariously to steep slopes directly above the canal and the river. So that Meilhan could still be 'sur Garonne', the engineers thoughtfully built a small tunnel beneath the canal to provide the villagers with access to the river below. With its riverside café and restaurant, Meilhan is a spectacularly beautiful place which alone would justify traversing those dreary miles of water out of Toulouse.

The glimpses of the Garonne that this section of the Lateral Canal affords make one marvel the more at the skill and courage of the men who once navigated it. At the time I saw it in September the river was a shallow, swift flowing stream far below. Despite its volume, it seemed dwarfed by the breadth of its bed and by the huge gravel shoals through which it meandered. Yet a flood gauge at Mas-d'Agenais left one in no doubt as to what this slumbering giant could do if it tried, for the top of the scale was almost on a level with the canal.

From Meilhan to Castets-en-Dorthe where the Canal ends in the tidal Garonne is 10½ miles with five locks, excluding those at Castets itself. What was a steep hill at Meilhan has become a cliff at Castets and on the brink of it the château and village are perched, looking down across a narrow strip of greensward to the canal, which here widens into a basin, and beyond it to the broad tidal river. At the western end of the basin are the two tidal locks. It is 35 miles (56 km) from these locks to the Pont de Pierre at Bordeaux which forms the official line of demarcation between inland and maritime navigation.

Although in theory this stretch of the river is navigable at any state of the tide by vessels drawing 6 ft 6 inches (2 m), a number of rocky shelves exist on the upper part of the river between Langon and Castets and it is said that at low water neaps the depth over these can be as little as 3 ft 7 inches (1·1 m). For this reason, coupled with the fact that the current in the river is at all times swift, craft usually leave Bordeaux four or five hours before high water so as to arrive with the tide at Castets. Similarly, traffic proceeding downstream moves on the ebb. The deep-water channel shifts about and is not buoyed or marked in any way so that strangers to the navigation usually very wisely follow in the wake of commercial traffic.

A certain number of powerful self-propelled barges loading 800 tons work on this stretch of the river alone, some carrying sand or gravel dredged from its bed and others hydrocarbons for the supply of ships and tanks in the Bordeaux area. But the canal supplies the bulk of the traffic, and the stranger may well wonder why its builders did not extend their Lateral Canal all the way to Bordeaux. For although this tidal river may hold no terrors for the professional boatman, it remains a hazardous passage allowing little margin for error or mechanical failure. Moreover, as already mentioned, it necessitates equipping canal craft with larger and more powerful engines than would be the case if they worked in still waters only.

We have now traversed this great inland waterway from the tideless blue waters of the Étang de Thau on the shores of the Mediterranean to an estuary scoured by powerful Atlantic tides. We have noted along the way the various improvements that the generations of engineers who succeeded Riquet have made to it. It is no disparagement of these successors to say that with one exception they have all been improvements of detail. The exception is the Lateral Canal, yet even here, if tradition speaks truly, the original conception was Riquet's. As for the Canal du Midi itself it is surely remarkable that after nearly 300 years it should still be serving its commercial function so substantially unchanged. Most remarkable of all is the fact that its key feature, that master-stroke of Riquet's genius, the feeder system from the river Sor and the Montagne Noire, still functions precisely as he planned it and is still adequate to meet today's needs. But for this there could be no traffic between the two seas.

What of the future? It is hoped eventually to reconstruct and modernise the entire canal route from the Garonne to the Rhône so that it can accept barges of 350 tonnes capacity, this being the minimum standard size for European inland waterways. This plan involves increasing the depth of water to 8 ft (2·6 m) and increasing the length of the locks from the present 100 ft (30·5 m) to 126 ft (38·5 m). So far, only the enlargement of the Garonne Lateral Canal has been authorised and the locks on the section between Castets and Agen have already been lengthened. In connection with this modernisation, the four locks at Agen, immediately west of the aqueduct, have been converted to oleo-hydraulic operation electrically controlled from a console in a lock-side cabin. Ironically enough, however, when we passed through these locks we found that there was a power failure and we had to resort to the emergency manual operation which proved to be extremely slow and tedious.

For service on this new Castets–Agen section a number of

existing barges have been lengthened by cutting them in half and inserting a new midships section. Some, if not all, of these midships sections have been provided by cutting up old barges at Sète, and many of the barges due for lengthening have passed through the Canal du Midi pushing their additional sections ahead of themselves.

The work of enlarging the second section of the Lateral Canal between Agen and Toulouse is already well advanced and it includes one experimental feature of great interest to all champions of inland water transport. This is the so-called 'water slope' which, when it is completed by, it is hoped, the end of 1972, will by-pass the five existing locks at Montech. This is an old idea which, in its present form, owes its origin to M. Jean Aubert, Inspecteur Général des Ponts et Chaussées. A model of this device to a scale of 1:10 was constructed at Vénissieux, near Lyons in 1967 and its results were sufficiently promising to justify a second full-scale experiment at Montech. It is necessary to emphasise the word experiment because to construct such an arrangement merely to supersede a flight of five single locks would otherwise scarcely be commercially justifiable.

The water slope itself consists of an inclined rectilinear concrete trough linking the lower and upper pounds of the canal, the waters of the upper pound being confined by a stop-gate. A triangular 'wedge' of water, in which the barge floats end-on, is pushed up the slope by a moving dam, called a shield, like a gigantic bulldozer blade. With this shield at the top of the slope, the water makes a level so that the stop gate can be opened for the passage of traffic from or to the water slope.

To propel the shield at Montech, two 1000 h.p. diesel-electric locomotives are to be used. These will be mounted

on giant pneumatic tyres and run on concrete tracks on
each side of the water slope on the model of the rubber-
tyred trains on the Paris Métro. The horizontal length of the
slope at Montech is 1575 yards (480 m) and the difference
of level overcome is 46 ft (14 m) so that the gradient of the
slope is 1 in 34·3 or 3 per cent.[1]

The two critical factors determining the gradient are the
maximum water pressure on the shield and the dimensions
of the barges using the slope. For it will be obvious that the
size and weight of the 'wedge' of water to be propelled is
determined by the length and laden draught of the vessel
which must float on it.

It is obvious also that the profile of the trough in cross-
section must be accurate and the seal round the shield
effective if undue loss of water is to be avoided. It is understood that rollers of synthetic rubber are to be used for
sealing. What steps have been taken to obviate possible
icing difficulties in winter are not known.

The two locomotives are due for delivery to Montech
during August 1972 which means that, if all goes well, the
first full-scale tests of the water slope should take place long
before these words appear in print. All inland waterway
advocates await their outcome with interest.

It is envisaged that when the enlargement of the Canal du
Midi is put in hand many of the acute bends, particularly on
the famous long pound, will have to be straightened out.
The lengthening of the locks will involve replacing the
present double and staircase locks by fewer and deeper
chambers. This naturally raises the question of the adequacy
of Riquet's feeder system to meet the needs of the enlarged
canal and whether the summit supply will have to be aug-

1. I am indebted to Mr David Edwards-May for this information.

mented. The answer is that the cost of electric power in France is so low that current engineering opinion holds that the cost of pumping back lockage water electrically is less than that of providing and maintaining elaborate gravity feeder systems. However logical this may seem, remembering the power failure at the locks of Agen, one questions the wisdom of making the efficient operation of an inland waterway wholly dependent on power supplied from a remote source.

Although finance for the modernisation of the Canal du Midi has not yet been authorised, it seems likely that ours may be the last generation to see it in its original state and I, for one, am certainly glad to have done so. For although the canal as we see it today owes much to the work of generations of French engineers from Vauban and Niquet in the seventeenth century to Magues and Simonneau in the 1850s, all these successors saw in it the reflection of Pierre Paul Riquet's unique genius and sought to emulate it. They aspired to make their own works worthy of that genius by endowing them with the same qualities of permanence and monumental grandeur. So it has come about that the unity of the whole is such that it could well be taken for the conception of a single presiding mind. That mind had decreed that the Canal du Midi should be something more than an heroic feat of engineering. Thanks to the unique privileges granted by Louis XIV to Riquet and his heirs, he deliberately set out to create what, in modern planner's jargon, would be called a linear park; the canal was to be the Riquet estate and he lost no opportunity to beautify it. Consequently no financial stringency, no engineering problem however intractable it might appear, was permitted to override aesthetic judgment on matters of landscaping or architecture. Hence Riquet's great canal has become an eloquent memorial of an age when the marriage of the arts and sciences, of beauty with utility, was taken for granted and their divorce inconceivable.

It is to be hoped that the modernisers of the Canal du Midi will respect its historical significance and strive to make their work worthy of it, but this is not to say that I am against modernisation as such. For any work of man's hands there must, sooner or later, come a time when it has either outlived, or can no longer fulfil, the purpose for which it was created. When this point is reached, either it must be embalmed as a museum piece, be adapted to serve some different purpose, or be modified so that it can continue to perform its original function. In any event, change becomes inevitable. From what I have learned of the character of Pierre Paul Riquet, Baron of Bonrepos, if faced with these three choices for the future of his canal, I am sure I know which one he would prefer.

[Appendix A]

ROYAL EDICT
For the Construction
of a Canal for the Communication of the Two Seas
Ocean and Mediterranean

Resolution of the Council of State and Letters Patent
in Interpretation thereof

Louis, by the Grace of God, King of France and of Navarre: to all those present and to come, GREETING. Whereas the proposal which has been made to Us for joining the Ocean and Mediterranean seas by a transnavigational Canal and for opening a new harbour on the Mediterranean on the coasts of Our Province of Languedoc, had seemed so extravagant to former centuries that even the boldest Princes and the Nations which have left to posterity the fairest evidence of their tireless labours, being daunted by the mightiness of the enterprise, and incredulous that it could be accomplished; nevertheless, as the boldest designs are those most worthy of men of high courage, and, provided they are prudently considered, they are usually successfully executed; so the repute of the undertaking of linking the two seas and the infinite advantages which were presented to Us as likely to benefit trade, convinced Us that it was a great enterprise of peace, well worthy of Our interest and care, capable of perpetuating for centuries to come its author's memory and emphasising the grandeur, the abundance and the happiness of Our reign. Indeed, We have seen that a connection between the two seas would enable all the nations of the world, as much as Our own subjects, to make in a few days through a canal route across the lands of

Our realm, and with little expense, a voyage which today can be undertaken only through the Straits of Gibraltar at the cost of great expense and much time, and with hazards of piracy and shipwreck. Consequently, with the object of making Commerce flourish in Our Realm by such substantial advantages, but nonetheless desirous of undertaking nothing except with the prospect of assured achievement; We have, after most thorough deliberation of the proposals laid before Us in favour of constructing a Canal that should join the two Seas, deputed Commissioners, drawn from the Three States of the said Province of Languedoc to go, together with the Commissioners presiding over the said States, jointly to the spot with persons experienced in the construction of the said Canal, so as to give Us their opinion on the feasibility of the venture. Which having been done by the said Commissioners with great circumspection and knowledge, they gave Us their advice on the possibility of carrying out the said proposals, and on the way and manner in which the said Canal could be effected and attempted and, in order to do so, to make as an experiment a small canal, cut out and conducted to the very same place where the construction of the large Canal is projected, this being so cleverly contrived and satisfactorily achieved by the care of Sieur de Riquet that We have every reason safely to foresee a most happy success. But, because a work of such magnitude cannot be undertaken without a very great expense, We have instructed our Council to examine the divers proposals made to Us to find funds without overburdening Our subjects in Our Provinces of Languedoc and Guienne, though they be the most justly bound to contribute thereto, because they will receive the first and greatest benefit therefrom; and We have determined upon those which appeared to Us the most supportable and the least harmful, in pursuance whereof it is needful that an appropriation be made.

Wherefore, the King orders that the construction of the Canal shall be undertaken in accordance with the estimate of the

Chevalier de Clerville, that the concern shall take hold of all necessary estate and money which shall be paid by His Majesty after valuation, that the lords of fiefs having jurisdiction over these lands will be indemnified, thereafter said lands to constitute a fief comprising the Canal, its feeders and banks, from the Garonne to the Mediterranean, including the diversion canal coming from the Montagne Noire to the Naurouze Stones. The owner of such a fief shall have the exclusive rights to build on the Canal banks a country seat, mills, docks for storing merchandise and houses for employees. The owner and his successors shall own this fief forever with exemption from duty (*taille*) and enjoy hunting and fishing rights. Furthermore, he shall enjoy the exclusive right to build boats for the carriage of merchandise, also to appoint judiciaries and twelve guards, bearing the King's livery, to apply their sentences. In order to ensure that its perpetual financing does not become a burden either to the King or to the Province, His Majesty determines the rates chargeable on goods carried on the Canal; and finally, orders the sale of functions and salt duties, applying the moneys coming therefrom to the cost of that construction.

WHEREFORE, and in consideration of all other matters urging Us thereunto, on the opinion of our Council, and from our own certain knowledge, and full power and Royal authority.

For such is our pleasure, and in order that this may be a matter well and firmly established forever We have ordered our Seal to be set upon these Presents.

GIVEN at SAINT-GERMAIN en LAYE in the month of October in this year of Grace 1666, and of our Reign the twenty-fourth,

LOUIS

By Order of the King:
PHELYPEAUX

Seen in Council:
COLBERT

[Appendix B]

OPERATING EXPENSES

The following annual figures were given to Huerne de Pommeuse in 1818 by M. Clauzade, Chief Engineer of the Canal du Midi, and were said by him to represent an average taken over the previous eighteen years.

	Francs
Provision of Water Supplies	75,000
Construction Works, excluding locks and cuts	60,000
Drainage Works	54,000
Construction work on locks and cuts	85,000
Maintenance of canal bed	80,000
Maintenance of 170 lock gates, to be renewed every 15 years approximately	40,000
Supply of Pozzolana, coming from St Peter of Rome, for cement	17,500
Machinery and Engines for the works	20,000
Maintenance of lodgings, sheds, etc.	15,000
Factories	3,000
Post Barges { Maintenance	9,600
Towing costs: 18 masters, 20 relays, 42 horses, 20 coachmen	36,600
Land Tax	48,000
Various Maintenance Expenses	12,000
Wages and Salaries	142,000
Pensions	13,000
TOTAL	**710,700**

[Appendix C]

ITINERARY

	Lock chambers	Kilometres
Toulouse:		
Port de l'Embouchure,		
Pont Jumeaux and		
Junction with Lateral Canal to:		
Toulouse:		
Bearnais Lock	1	1
Minimes Double Lock	3	2
Matabiau Lock	4	3
Bayard Double Lock	6	3
Port St Sauveur		5
Castanet Double Lock	8	15
Vic Lock	9	17
Montgiscard Double Lock	11	24
Aigues-Vives Double Lock	13	28
Sanglier Double Lock	15	29
Négra Lock	16	33
Laval Double Lock	18	37
Gardouch Lock and Port	19	38
Renneville Lock	20	43
Encassan Double Lock	22	45
Emborrel Lock	23	47
Ocean Summit Lock, Naurouze	24	51
Le Ségala		53
Mediterranean Summit Lock	25	56
Roc Double Lock	30	58
Domergue Lock	31	59

	Lock chambers	Kilometres
Laplanque Lock	32	60
Castelnaudary:		
Port		64
Grand Bassin		65
St Roch Quadruple Staircase Lock	36	65
Gay Double Lock	38	67
Vivier Triple Staircase Lock	41	68
Guilhermin Lock	42	69
St Sernin Lock	43	69
Guerre Lock	44	70
Peyruque Lock	45	71
Criminelle Lock	46	72
Tréboul Lock	47	73
Villepinte Lock	48	77
Sauzens Lock	49	79
Bram Lock and Port	50	80
Béteille Lock	51	85
Villesèque Lock	52	93
Lalande Double Lock	54	98
Herminis Lock	55	98
Ladouce Lock	56	99
Pouillariès, junction of old Fresquel Line		101
Carcassonne, Port and Lock	57	105
St Jean Lock	58	107
Fresquel:		
Aqueduct		108
Double Lock	60	108
Junction of old Fresquel Line		108
Lock	61	108
Evêque Lock	62	112
Villedubert Lock	63	113
Orbiel Aqueduct		116

	Lock chambers	Kilometres
Trèbes:		
Port		117
Triple Staircase Lock	66	118
Marseillette Lock	67	127
Fonfile Triple Staircase Lock	70	130
Saint Martin Double Lock	72	131
Aiguille Double Lock	74	133
Puicheric Double Lock	76	136
Laredorte, Port		140
Jouarres Lock	77	142
Homps:		
Port		145
Lock	78	146
Ognon Double Lock and Aqueduct	80	147
Garde d'Ognon flood gates		
(*demi écluse*)		147
Pechlaurier Double Lock	82	149
Argens Lock and beginning of		
long pound	83	152
Roubia, Port		154
Paraza, Port		157
Répudre Aqueduct		158
Ventenac, Port		160
Le Somail, Port		165
Cesse Aqueduct		167
Junction with Narbonne Branch Canal		168
Argeliers, Port		
Pigasse Bridge, cottage and stop planks		178
Capestang, Port		188
Poilhes, Port		194
Malpas tunnel		198
Head of Fonserannes Octuple Staircase:		
Lock and end of long pound		206

	Lock chambers	Kilometres
Fonserannes:		
Junction of old Orb Line at		
7th chamber	89	206
Béziers:		
Orb Aqueduct		207
Orb Double Lock	91	208
Port		208
Double Lock, junction of old		
Orb line	93	208
Ariège Lock	94	212
Villeneuve Lock	95	213
Portiragnes Lock	96	218
Libron Flood Gates		225
Agde:		
Port		231
Round Lock, Junction of Agde Branch	97	231
Hérault River Crossing		232
Prades Flood Lock	98	232
Bagnas Lock	99	235
Onglous, Port		239
Junction with Étang de Thau		240

Note:

The following is a list of Riquet's original locks which have either fallen into disuse or have been superseded and become ruinous:

Toulouse:
    Garonne Double Lock   2
Old Fresquel Line:
    Foucaud triple staircase lock   3
    Villaudy double lock   2
Old Orb Line:
    Fonserannes, 7th and 8th chambers   2
    Notre Dame double lock   2

                        TOTAL   11

Because the Prades flood lock was built subsequently, there were originally 101 lock chambers on the canal from the Garonne to the Étang de Thau, 26 on the western side of the summit and 75 on the eastern.

[Appendix D]

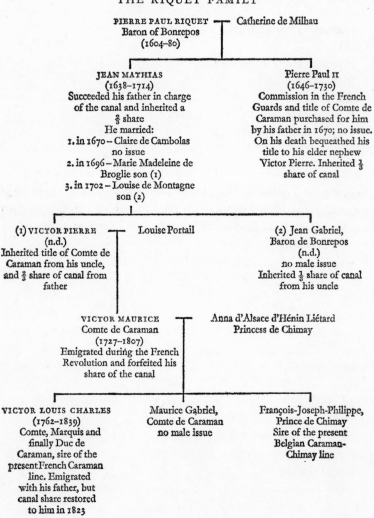

THE RIQUET FAMILY

PIERRE PAUL RIQUET ── Catherine de Milhau
Baron of Bonrepos
(1604–80)

JEAN MATHIAS
(1638–1714)
Succeeded his father in charge
of the canal and inherited a
⅔ share
He married:
1. in 1670 – Claire de Cambolas
no issue
2. in 1696 – Marie Madeleine de
Broglie son (1)
3. in 1702 – Louise de Montagne
son (2)

Pierre Paul II
(1646–1730)
Commission in the French
Guards and title of Comte de
Caraman purchased for him
by his father in 1670; no issue.
On his death bequeathed his
title to his elder nephew
Victor Pierre. Inherited ⅓
share of canal

(1) **VICTOR PIERRE** ── Louise Portail
(n.d.)
Inherited title of Comte de
Caraman from his uncle,
and ⅔ share of canal from
father

(2) Jean Gabriel,
Baron de Bonrepos
(n.d.)
no male issue
Inherited ⅓ share of canal
from his uncle

VICTOR MAURICE ── Anna d'Alsace d'Hénin Liétard
Comte de Caraman   Princess de Chimay
(1727–1807)
Emigrated during the French
Revolution and forfeited his
share of the canal

VICTOR LOUIS CHARLES
(1762–1839)
Comte, Marquis and
finally Duc de
Caraman, sire of the
present French Caraman
line. Emigrated
with his father, but
canal share restored
to him in 1823

Maurice Gabriel,
Comte de Caraman
no male issue

François-Joseph-Philippe,
Prince de Chimay
Sire of the present
Belgian Caraman-
Chimay line

[Acknowledgements]

The preparation of this book was only made possible by the kindness and help of a number of people on both sides of the Channel. First of all I have to thank my friends Michael Streat, of Blue Line Cruisers (France) Ltd, and Sir Robert Grant-Ferris for making possible the 'on-the-spot' part of the enterprise. For it was the former who was responsible for organising my first, eastbound voyage through the canal from Castelnaudary to Marseillan, while it was entirely due to Sir Robert's kindness and hospitality that I was able to make a second and ever-memorable westbound voyage from Agde to Castets-en-Dorthe on board his motor yacht *Melita*.

For making good my inadequate command of the French language I am very deeply indebted to my old friend and collaborator M. Henri Delgove of Le Mans who most kindly agreed to act as my research assistant. Not only did he do a great deal of translation work for me, but he also acted as my intermediary, receiving and translating my numerous practical queries, forwarding them to the appropriate authorities, and then translating their answers. Most of these queries were addressed to M. André Pfaff, until very recently Ingénieur-en-Chef des Ponts et Chaussées, Service Navigation Midi-Garonne, Toulouse. M. Pfaff was an extremely busy man, yet he found time to answer very fully the numerous questions I put to him concerning the history and the practical engineering of the canal. I feel I cannot thank him enough for his kindness and courtesy in supplying me with so much invaluable information.

I am also very grateful to Terry Adolph, Manager of Blue Line Cruisers (France) Ltd at Castelnaudary, for the enthusiastic way he pursued various lines of inquiry on my behalf and also to those who supplied him with information, notably the District Engineer, M. Lorblanchet, and M. La Croix. M. La Croix is a retired lock keeper from Domergue Lock whose father, grandmother and great-grandfather were all born in the Domergue lock cottage. He supplied valuable factual information about the changeover from horse towage to motorised barges in the early 1920s.

On this side of the Channel I am grateful to Professor A. W. Skempton, Dr Stephen Johnson, Christopher Marsh, Charles Hadfield and Dr Roger Pilkington for their help and advice; also to the Librarian of the Institution of Civil Engineers for providing research facilities. Finally, I wish to express my thanks to Mlle Mireille Lefevre, Roger Crabtree and my wife for helping me to wrestle with early French texts.

L.T.C.R.

[Bibliography]

French

Andreossy, Général F., *Histoire du Canal du Midi ou Canal de Languedoc*. Paris: L'Imprimerie de Crapelet, 1804.
A very fully detailed history of the canal. Its only fault is that the author unduly magnifies his great-grandfather's part in its construction at the expense of Riquet. There is a separate volume of plates from which a number of the illustrations in this book were selected.

Girou, Jean, *Nòstre Riquet*, Toulouse: College d'Occitanie, 1968.
The only modern biography of Riquet and as eulogistic about its subject as Général Andreossy's History is disparaging. Well documented with many lengthy verbatim extracts from the Riquet/Colbert correspondence. Includes a good bibliography.

De La Lande, *Des Canaux de Navigation et Specialement du Canal du Midi*, Paris: Chez la Veuve Desaint, 1778.
Includes a history and description of the canal which was approved by Riquet's descendants; also a few maps and plates.

Pfaff, André, 'Les Voies Navigables du Sud-Ouest', in *Regards sur la France* sp. ed., 1966.
A useful description of the canal and its traffic in 1965 by the recently retired Chief Engineer.

De Pommeuse, H., *Des Canaux Navigables*, Paris: L'Imprimerie de Huzard-Courcier, 1822.
English and French canals described and compared. There is an 82-page section on the Canal du Midi which includes much

factual information supplied to the author by M. Clauzade, the then chief engineer. Contains plates and maps.

English

Pilkington, Roger, *Small Boat through Southern France*. London: Macmillan, 1965.

An entertaining account of the author's voyage in his boat *Commodore* from Beaucaire on the Rhône by the canals and the Garonne to Bordeaux. It includes much local colour and history and can therefore be recommended as a complement to this book.

Skempton, A. W., Canals and River Navigations, *A History of Technology*, vol. iii, Oxford: The Clarendon Press, 1957.

Much of the material in the first chapter of this book is drawn from this admirably concise account of the development of canal and river navigation engineering in Europe before 1750.

Smith, Norman, *A History of Dams*, London: Peter Davies, 1971.

An excellent and most valuable world-wide study of a neglected but fascinating subject. It includes descriptions of the St Ferreol and Lampy dams and so helped me to set them in their true historical perspective.

[Index]